Extraterrestrials

Stan Gable
May 1988

EXTRATERRESTRIALS

SCIENCE AND ALIEN INTELLIGENCE

Edited by

EDWARD REGIS JR

Howard University, Washington DC

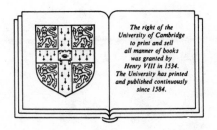

The right of the
University of Cambridge
to print and sell
all manner of books
was granted by
Henry VIII in 1534.
The University has printed
and published continuously
since 1584.

CAMBRIDGE UNIVERSITY PRESS

Cambridge

New York New Rochelle Melbourne Sydney

Published by the Press Syndicate of the University of Cambridge
The Pitt Building, Trumpington Street, Cambridge CB2 1RP
32 East 57th Street, New York, NY 10022, USA
10 Stamford Road, Oakleigh, Melbourne 3166, Australia

First published 1985
First paperback edition 1987
Reprinted 1987 (twice)

Printed in the United States of America

Library of Congress catalog card number: 84–23781

British Library Cataloguing in Publication Data
Extraterrestrials: science and alien intelligence.
1. Life on other planets
I. Regis, Edward
574.999   QB54
ISBN 0 521 26227 5 hard covers
ISBN 0 521 34852 8 paperback

CONTENTS

	Contributors	vi
	Preface	ix
1	**Overview**	1
	Lewis White Beck: Extraterrestrial intelligent life	3
2	**Existence and nature of extraterrestrial intelligence**	19
	Ernst Mayr: The probability of extraterrestrial intelligent life	23
	David M. Raup: ETI without intelligence	31
	Michael Ruse: Is rape wrong on Andromeda? An introduction to extraterrestrial evolution, science, and morality	43
3	**Extraterrestrial epistemology**	79
	Nicholas Rescher: Extraterrestrial science	83
	Marvin Minsky: Why intelligent aliens will be intelligible	117
4	**Where are they?**	129
	Frank J. Tipler: Extraterrestrial intelligent beings do not exist	133
	Carl Sagan and William I. Newman: The solipsist approach to extraterrestrial intelligence	151
5	**Detectability and decipherability**	163
	Jill Tarter: Searching for extraterrestrials	167
	Cipher A. Deavours: Extraterrestrial communication: A cryptologic perspective	201
	Hans Freudenthal: Excerpts from *LINCOS: Design of a language for cosmic intercourse*	215
6	**Meaning and consequences of contact**	229
	Edward Regis Jr: SETI debunked	231
	Jan Narveson: Martians and morals: How to treat an alien	245
	Robert Nozick: R.S.V.P. – A story	267
	Index	275

CONTRIBUTORS

Lewis White Beck:
Department of Philosophy,
University of Rochester,
Rochester, New York 14627, USA

Cipher A. Deavours:
Department of Mathematics,
Kean College of New Jersey,
Union, New Jersey 07083, USA

Hans Freudenthal:
Frans Schubertstraat 44,
3533GW Utrecht,
The Netherlands

Ernst Mayr:
Museum of Comparative Zoology,
Harvard University,
Cambridge, Massachusetts 02138, USA

Marvin Minsky:
Artificial Intelligence Laboratory,
Massachusetts Institute of Technology,
Cambridge, Massachusetts 02139, USA

Jan Narveson:
Department of Philosophy,
University of Waterloo,
Waterloo, Ontario,
Canada N2L 3G1

William I. Newman:
Department of Earth and Space Sciences,
University of California,
Los Angeles,
Los Angeles, California 90024, USA

Robert Nozick:
Department of Philosophy,
Harvard University,
Cambridge, Massachusetts 02138, USA

David M. Raup:
Department of Geophysical Sciences,
University of Chicago,
Chicago, Illinois 60637, USA

Edward Regis Jr:
Department of Philosophy,
Howard University,
Washington DC 20059, USA

Nicholas Rescher:
Department of Philosophy,
University of Pittsburgh,
Pittsburgh, Pennsylvania 15260, USA

Michael Ruse:
Departments of History and Philosophy,
University of Guelph,
Ontario,
Canada N1G 2W1

Carl Sagan:
Laboratory of Planetary Studies,
Cornell University,
Ithaca, New York 14853, USA

Jill C. Tarter:
Space Sciences Laboratory,
University of California,
Berkeley, California 94720, USA
and
NASA Ames Research Center,
Moffett Field,
California 94035, USA

Frank Tipler:
Department of Mathematics,
Tulane University,
New Orleans, Louisana 70118, USA

TO CARL SAGAN
AND
LEWIS WHITE BECK

PREFACE

Are extraterrestrials imaginary creatures, the otherworldly equivalent of ghosts, demons, or the Loch Ness Monster? Or are they really and truly out there, real things in the real world even as you and I? If they do exist, what are our prospects for contacting them, or for them contacting us? How would such contact be made? And would messages from truly alien beings be intelligible to us?

Speculation about ETI (extraterrestrial intelligence) is popular these days, both in mass culture and in science, but belief in extraterrestrials goes back to ancient times. It is only recently, though, that this belief has been capable of being verified or falsified by experimental test. If we should one day find out that intelligent aliens do exist, what will the discovery mean? Will it irretrievably alter our view of ourselves, or will human life go on pretty much as it had before?

Recently, the physical sciences have begun to take these questions very seriously. Chemists have pondered the requirements for the origin of life on other worlds. Evolutionists have considered whether and how that life might develop. Physicists have calculated the rates at which interstellar cultures might populate the galaxy. Also, philosophers have tried to say something about the significance, the meaning in the larger scheme of things, that contact with intelligent aliens would have.

Often enough, the thinking of these scholars is communicated to their peers only at conferences or in scattered technical journals, making it difficult for academics in other disciplines, or for the layman, to hear what these people have to say. The purpose of this volume is to present in an easily accessible form the best and most recent original thought on these questions from some of the most distinguished scientists and philosophers of our generation.

I would like to acknowledge with thanks the help given during the preparation of this book by the staff of Cambridge University Press and Mrs Sandi Irvine. I especially want to thank my wife, Pamela Regis, for preparing this volume's index.

Laramie, Wyoming   E. R. Jr
July 1984

PART 1

Overview

An overview of the many scientific and philosophical issues raised by the search for extraterrestrial intelligence (SETI) is given by Lewis White Beck, Professor of Philosophy at the University of Rochester. In 1971, as President of the Eastern Division of the American Philosophical Association, Beck delivered an address entitled 'Extraterrestrial Intelligent Life.' The address contains a history of belief in ETI and explains the major obstacles to the development of intelligent life on other worlds. Additionally, it presents an account of the barriers to the possibility of communication with aliens, and projects what would be the likely consequences of eventual success. A thoughtful, authoritative, and clear account of these issues, it is probably the best single paper ever written on the hopes and hurdles of SETI. For this reason, Beck's presidential address is reprinted here.

Extraterrestrial intelligent life*

LEWIS WHITE BECK

I must confess that I have a singular reason for being gratified at the privilege of delivering this paper as a presidential address – if it were not the presidential address I doubt that it would be accepted by the Program Committee. Our Association is not hospitable to cosmological speculations. I can avail myself only of presidential license in asking you to consider a perennial theme in philosophy which is neglected by philosophers at a time when it is most cherished by scientists. Many eminent philosophers – among others Aristotle,[1] Nicholas of Cusa, Giordano Bruno, Gassendi,[2] Locke,[3] Lambert,[4] Kant,[5] and William Whewell[6] – have believed that there is extraterrestrial life; yet I know of only one or two living professional philosophers writing in English[7] who have even discussed the question. This is unfortunate, since many of the problems our scientific colleagues are raising, such as those of the criteria of life, mind, intelligence, and language, and the future viability of our civilization, are problems about which we philosophers have much to say. There are new sciences like exobiology whose foundations are in need of philosophical scrutiny. When the National Academy of Sciences explicitly calls attention to the philosophical dimensions and ramifications of the problem,[8] it seems to me we philosophers should relax our ban on cosmological speculation and think about possible worlds that may actually exist.

I

The belief that there are animate and superhuman beings inhabiting heavenly bodies, or that heavenly bodies are themselves animate and

* Presidential address delivered before the Sixty-eighth Annual Eastern Meeting of the American Philosophical Association in New York City, December 28, 1971.

conscious – it is often difficult to know which – was very widespread in the ancient Mediterranean world. When many of the ancient Greek philosophers[9] asserted that the moon, planets, and stars are inhabited, they probably did so without great speculative venturesomeness; it might have been more idiosyncratic to have denied so commonplace a view. Though Aristotle believed that the moon is inhabited, the Aristotelian teaching of the uniqueness of the world put an end to easy assumptions.

The two ancient writings of most interest for our topic are those of Lucretius and Plutarch. That two writers having as little in common as they did came by very different arguments to much the same conclusion suggests that we have here to do with a myth or archetypal idea. Like the idea of God, to which it is not unrelated, the belief that we are not alone appeals on a prephilosophical level where men as different as Plutarch and Lucretius are at one. These two writers set the ground rules for all future speculation on extraterrestrial life. Let it be granted that they did not know even what little we know about the answers to their questions; but they knew the right questions to ask, and their questions and some of their answers have been repeated again and again.

Plutarch's thesis in *The Face that is in the Orb of the Moon* is that there are, or at least may be, men in the moon. He reaches this conclusion from four premises. (1) The earth has no privileged position in the universe (925 E–926 C). (2) The earth and heavenly bodies are not where they should be in accordance with the doctrine of natural positions and movements; hence the matter of the universe has been distributed by an intelligence working by design (926 F–927 F, 928 C–D). (3) The moon is sufficiently like the earth to support life (937 C–D). (4) If the moon did not support life, it would exist to no purpose, and this is inconsistent with the premise of intelligent design (937 E). From these it is supposed to follow that there are living beings on the moon.

Lucretius' *De rerum natura* was written before Plutarch's dialogue, but it may best be considered as if it were a criticism of it, for Lucretian reactions take place repeatedly against a long line of Plutarchean arguments. Lucretius accepts only two of the four premises of Plutarch's argument, namely the first and the third. There is an infinity of empty space with atoms jostling about, and here and there are 'gatherings of matter' (II, 1044 ff) some of which are other worlds with their own skies and races of men and beasts (II, 1067 ff) which have originated in natural ways, have undergone divergent natural selection in diverse environments, and have developed diverse civilizations (V, 774 ff). There is no design; the celestial bodies do not exist for the sake of their inhabitants; their inhabitants exist by chance. Lucretius does not stop with the paltry question of whether the moon is sufficiently earth-like to support life; he says that there must be innumerable worlds in all degrees of likeness and unlikeness, and therefore there must be worlds which have inhabitants some like us and some unlike.

These speculations by Plutarch and Lucretius set up the ways of arguing for two thousand years. New astronomical and biological information has been repeatedly put into these two forms. The Plutarchean argument became a part of the natural theology of Christianity after the Copernican revolution; the Lucretian argument came into its own after the Darwinian.

But no successful elaboration of these promising beginnings, which we cannot help admiring even today, was made for fourteen centuries. St Augustine exercised his great authority against the plurality of worlds;[10] in the eighth century a bishop was removed from office for affirming plurality; and it was formally declared heretical in the eleventh.[11] When the teachings of Aristotle were introduced in Paris in the thirteenth century, added weight was given to the dogma of the uniqueness of the world, so much weight indeed that it threatened the dogma of the power of God to go against the natural Aristotelian order. So in the Condemnation of 1277 it was forbidden to teach that God could not have created a plurality of worlds (Article 27). This was not done in defense of the thesis of an actual plurality, but only in defense of the omnipotence of God. Both Albertus Magnus and St Thomas Aquinas had already held it to be in the power of God to create more worlds than one, but had denied, on good Platonic and Aristotelian ground, that He had in fact done so.[12]

While the normal form of teleology in the Middle Ages had been man-centered or God-centered, the doctrine of the plurality of worlds based upon Copernicus broadened the teleological framework. While every part of the universe was designed for a purpose or at least had a purpose in the organic and spiritual unity of nature – no one denied that – man and our own earth were no longer seen as the sole purpose. When it was decided that the earth is a planet and the sun a star, the pervasive teleological conviction implied that there are other living beings for whose benefit the planets and stars exist. Nicholas of Cusa drew this Plutarchean inference even before Copernicus. Both he and Bruno rejected the craft-handiwork conception of a teleology of design in favor of the animistic neo-Platonic principle of the plenitude of being, holding that the perfection of each heavenly body requires that it support spiritual life. '[Other worlds] are not required for the perfection and subsistence of our own world, but . . . for the subsistence and perfection of our universe itself an infinity of worlds is necessary,' wrote Bruno.[13]

By the seventeenth and eighteenth centuries, the belief had become a commonplace. It was supported sometimes by neo-Platonic, sometimes by deistic, and occasionally by materialistic arguments which followed the Plutarchean or Lucretian paradigms.[14] The most important sources of the received opinion were works of Thomas Wilkins,[15] Christiaan Huygens,[16] and Bernard de Fontenelle.[17] While Montaigne,[18] Milton,[19] and Pope,[20] urged men to give more thought to mundane problems, others like Campanella,[21] Swift,[22] and Voltaire[23] followed Lucian[24] in using the imaginary inhabitants of other worlds as critics of, or as models for condemning and attempting to rectify, the follies of mankind. Throughout Europe astrobiology was an important part of the 'evidences of Christianity;' only a few men, such as Thomas Paine,[25] thought that the plurality of worlds rendered Christianity 'little and ridiculous, and scatters it in the mind like feathers in the air.' But on one point all were agreed: the universe was full of life, man was not alone.

The first detailed scientific criticism of the doctrine appeared in an anonymous work published in 1854 by William Whewell, *The Plurality of Worlds*. This book, by the eminent historian and philosopher of science well acquainted with the latest advances in geology, biology, and astronomy, is a criticism of the classical English formulation of pluralism as an evidence of Christianity, the writings of Bishop Thomas Chalmers.[26] It is also a silent renunciation – perhaps this explains its anonymity – of conjectures about extraterrestrial life which he had permitted himself in his Bridgewater Treatise[27] twenty years earlier.

From his imposing astronomical and geological learning Whewell inferred that the universe has only a small number of planetary systems and that the earth is uniquely able to support life.[28] He concluded:

> 'The belief that other planets, as well as our own, are the seats of habitation of living beings has been entertained in general, not in consequence of physical reasons, but in spite of physical reasons; and because there were conceived to be other reasons, of another kind, theological and philosophical, for such a belief.'[29]

These 'other reasons,' however, no longer existed for Whewell – not because he doubted the validity of arguments from design but because he held that arguments from design pointed to the opposite conclusion. The world would be imperfectly designed if the incarnation of Christ were only terrestrial while there were unredeemable souls elsewhere in the universe.[30]

Whewell's book caused a great stir in Victorian theological and astronomical circles and was severely criticized.[31] Hume's objection to the argument from design was not repeated by Whewell, but he was arguing as if he remembered Hume's words:

'The religious hypothesis . . . must be considered only as a particular method for accounting for the visible phenomena of the universe; but no just reasoner will ever presume to infer from it any single fact, or alter or add to the phenomena, any single particular.'[32]

While one might argue from extraterrestrial life to the purposive design of parts of the universe indifferent to man, Hume would forbid us to argue from the assumption of pervasive design to the existence of unobserved life. But I do not think anyone remembered Hume at this juncture; the argument from design was threatened much closer to home. It was considered more important to preserve the argument from design against Darwin, Huxley, and Wallace[33] to show that man is uniquely the purpose of terrestrial arrangements than it was to use the argument to show that he is *not* unique in the cosmic order.

There were soon two new scientific considerations which strengthened Whewell's case. The first was the refutation of the theory of spontaneous generation. Lucretius had believed that from the primal aggregations of matter life would spontaneously arise; now it seemed certain, on the best experimental evidence, that it would not. The second was the objections raised against the Kant–Laplace nebular hypothesis about the origin of the solar system. As long as this theory had been accepted, there had been ground for believing that all stars have planets. By the end of the century the evidence turned against it, and it was replaced in astronomical orthodoxy by the explanation of solar systems as results of close approaches of stars to each other. Since such approaches can occur only seldom, planetary systems and hence life must be extremely rare.

With the rejection of both the Plutarchean premise of design and the Lucretian premises of innumerable worlds and the natural generation of life, the ancient doctrine temporarily disappeared from both astronomy and theology; the 'canals' of Mars had only a *succès de scandale*. Yet now once again we find the belief widespread, embraced by eminent astronomers and biologists and supported by taxpayers. Old ideas like this never die. When one argument fails, another will be found. Now it is an inference from non-anthropocentric naturalism. This Lucretian thought, filled out by the work of Copernicus, Darwin, and Marx,[34] provides the philosophical basis for contemporary exobiology. But it is not the only one which motivates it, as we shall see.

II

The contemporary Lucretian argument stands on two legs, one astronomical and one biological. The astronomical is the hypothesis that the earth is not anomalous. Once again it seems that the occurrence of planetary systems can best be explained by a modification of the Kant–Laplace hypothesis, the implication being that many suns support families of planets. While the problem of the infinity of the universe is not as clear to us as it appeared to Lucretius, it seems that the universe gets larger with every advance in astrophysics. There are 10^{21} stars in the observable universe, and this is quite enough for highly improbable events to have occurred many times. Even if only one out of a million stars has at least one planet, there are 10^{15} planets in the universe. Actually the probability derived from astronomical theory and a few delicate observations (e.g. of Barnard's Star) is much higher than one out of a million, so in all probability the number of planetary systems is much larger than 10^{15}.

The biological premise is likewise the rehabilitation of a theory rejected in the nineteenth century. Once again it is believed that life arises spontaneously when conditions are 'right.' It was conjectured what the conditions were on the primitive earth, and then it was found in the laboratory that under these conditions organic molecules necessary for life are produced. One of the conditions we cannot replicate in the laboratory is their duration for billions of years, during which these compounds might polymerize and associate into simple living sytems from which the evolutionary process might begin. But according to strong scientific dogma this did occur on earth, and under like conditions elsewhere it is highly probable that the same process occurs there too.

But how like do these conditions have to be? No one knows, for at most only one case is known, and there is no well-established theory of the transition of complex organic molecules to simple living systems from which the requisite degree of similarity and the probabilities might be deduced. If only one sun in a million has a planet, and only one planet in a million supports life, there are a thousand million abodes of life in the universe. This line of thought, however dear to popular science writers, implies nothing about the actual universe unless we know whether the second of these fractions – one out of a million – is of the right order of magnitude. It is, to be sure, a small fraction, but for all we know it may be much too large. The chemistry underlying exobiology now provides only conjectures about what may perhaps be the necessary conditions of life; it says nothing about the sufficient conditions, or how probable is their occurrence.

The connecting link between the new astronomical and new biochemical arguments lies in recent empirical discoveries. There are organic materials in meteorites and in interstellar space, and the gases occluded in moon-rocks suggest that the conditions *necessary* for the origin of life were not confined to the primitive earth. But absolutely nothing is known of what the *sufficient* conditions are, or how pervasive or local they may be.

No one now expects to find advanced life elsewhere in the solar system. But in a few years we may know whether there is primitive life on Mars. A microorganism on Mars will convert 'the miracle of life' on earth into a 'mere statistic.' It will contribute markedly to the argument that advanced life is present outside the solar system.

III

Space travel, or even the sending of instrumental probes, to other solar systems is so far beyond human reach that it is not worth while discussing at a sober philosophical cocktail party. Nor do I think it promising to hope or fear that we will get evidence of the existence of superior beings by their visiting us. The only even moderately realistic hope for evidence lies in receiving and interpreting signals from extraterrestrial societies.[35] The technology required presents no insurmountable obstacles; what stands in the way of using it is human unimaginativeness and impatience and the instability of human civilization.

In the idea of interstellar communication, however, all the anthropomorphism so painfully eliminated with the Plutarchian argument insidiously reappears. To entertain hope of such converse requires that we believe that a pattern of evolution like ours, from simple organisms to advanced civilization, has been repeated within signalling distance and synchronously with our own development. It requires that the citizens of heavenly cities be sufficiently like us to reciprocate our curiosity and to take the same measures we would take to signal to them, and sufficiently unlike us to have managed a technological project which probably exceeds our resources of curiosity, patience, and stability. All of these assumptions are highly speculative, but we must make them or else give up the game.

We have to suppose that these creatures are enough like us to make communication from them intelligible to us. Even the assumption that their biochemical base is more like ours than we have any reason to believe[36] does not warrant the belief that the course of evolution will have gone in the same direction.[37] We do not even know whether conscious intelligence, at least in its higher forms, is in the long run biologically advantageous even on earth, let alone elsewhere. Speaking against this optimistic assumption is the fact that the only species on earth which prides itself on its intelligence is the only one with the intelligence necessary, and possibly sufficient, to render itself extinct tomorrow. We do not know whether the development of human-like species leads to species-suicide here or elsewhere; we do not know whether evolution elsewhere is likely to be progressive in terrestrial terms or not. We cannot assume that our evolution is both typical and non-lethal for any other reason than that not to assume it puts a sudden end to our research.

Even if we assume a common neurological base and evolutionary history, the cultural and technological aspects of human life are so loosely determined biologically that we must acknowledge the probability of widely divergent and perhaps unrecognizable manifestations of communal intelligence. We see this even on earth; why not expect it in the heavens? But everything that makes exotic extraterrestrial societies different from us reduces the probability of their disclosure. The feasible methods of interstellar communication are filters which will keep out evidence of the existence of beings whose logic, grammar, and technology, if they have such, are radically different from our own. To hope for evidence of living beings with radically different technologies based upon exotic sciences and logics is in principle vain. To believe that there are societies elsewhere bent upon and capable of communicating with us is not only to be anthropomorphic; even worse, it is to believe that civilizations elsewhere are like *one* civilization that has existed on only a small portion of this earth for only a few hundred years.

But for the sake of getting on with our work, let us grant that all these unlikely conditions may be fulfilled somewhere. But when? We run into the problem of synchrony. In order to communicate with each other, two planetary populations have to be alike at the same time.[38] One limiting factor is the density of civilizations in space, but an equally severe constraint is their density in time. The life-time of civilizations determines the probability of their having overlapping durations. Given optimistic estimates of their absolute densities in space (say two within a hundred light years of each other), the average life-time if they are to have signal-competency at the same time must be of the order of 10 million years.[39] In that length of time, biological change will predominate over cultural evolution. Suppose, however, that species and civilizations can survive long enough to overlap to a significant degree. We, being a 'young' technological civilization, must expect any other civilization with which we establish contact to be much older, technologically more advanced, and socially more stable than we are. But then we run into another limiting factor. The longevity of a technology is much less than that of a civilization. Out of perhaps a half million years of human life, we have had a radiomagnetic competency for fifty years, and fifty years from now it may be obsolete even if we still exist. Simultaneously existing civilizations are unlikely at any moment to possess compatible technologies of communication.

Thus our conjectures concerning history, especially the future history, of mankind are important variables in estimating our chances of finding out if we are alone. One has to be exceedingly optimistic about the future of mankind to be even moderately confident of getting an answer to our old question. Personally I do not feel such optimism, but let me talk a bit as if I did.

IV

Let us suppose that radio astronomers aiming their dish at a nearby star pick up a modulated radio signal that seems to have come from a planet. What will we do? How will we read it?

In the absence of collateral information, the necessary but not sufficient condition for knowing that something is a message is to know what it says.[40] Since the necessary collateral information will be lacking, we must guess that it is a message, guess what it says, and then try to see if the signal can convey that message. Since we cannot know what encipherment has been used, we must make conjectures about that too on an anthropomorphic model. We must ask what message we would send and what encipherment we would use if we were they. Plausible candidates are: a binary encipherment of the series of prime numbers, the expansion of e, or simple arithmetic truths which will contextually exhibit logical constants and operators.[41] It should be possible to match signals against such paradigms and look for a fit which would provide the necessary condition for deciding that a message had been received. But it would not be a sufficient condition, for the signal could have been produced by natural processes. To find out if it is an artifact of intelligence, we would follow Descartes' teaching that a machine can 'speak' but cannot 'discourse.' We would try to answer the putative message by continuing it and waiting the appropriate number of years or decades or centuries to see if the series is correctly continued still further.

If it is, we shall have as much proof as we can reasonably hope for. But there are an infinite number of rules by which a finite series of marks can be continued. Human imagination must limit the number of rules to be tried; and it is precisely the fact that this imagination is human which may be fatal to our hopes. If there is a sender, he may not recognize our return signal as an intelligent response, and we will have received a message without ever knowing it.

Let us suppose, however, that we pass the first test and establish the existence of an intelligent sender. That will be much, a very great deal, but not all we want. We want a vocabulary of denotative words. Professor Quine[42] has made us familiar with a problem we meet here in extreme form. For us to know that a string of marks is in a language, there must be observable objects or actions with which we know it to be correlated. But we do not know what facts extraterrestrial messages are to be about, since we do not know what kinds of things are up there to be talked about. We do not expect them to talk about rabbits as we do, but we do not even know whether 'undetached parts of rabbits'[43] are there to be talked about in strange ways.

Two means of providing the denotative components have been proposed. The messages might contain ostensive references to something we both can see, e.g. variations in the intensity of the sun.[44] By scanning all likely fields of empirical information available to both of us we could try to match the putatively denotative content of the messages with them.

Another means is superficially more promising: that they use television signals and supply us with a vocabulary.[45] This is not technically impossible, provided we use the word 'television' loosely. But there is an epistemological barrier which may be insurmountable. We do not know what things are up there and do not know how they will make themselves known to the sender. Therefore we have no way of knowing whether the pictorial array we achieve will correspond to any extraterrestrial fact. If their sensory channels are different from ours, as may well be the case, what is image for them will be snow for us. Only if the things they think worth talking about look to them in much the same way things we are interested in look to us can we correctly believe that we have received the picture they sent.[46]

V

I have emphasized, but I do not believe I have exaggerated, some difficulties in establishing the existence of intelligent extraterrestrial life. I have emphasized them because most of those now writing on the problem seem to me to minimize them. They have invariably assigned favorable probabilities to unknown but limiting conditions. Though they have insistently warned against the dangers of anthropomorphism, their models have been inescapably anthropomorphic. Their assumptions have not been made irresponsibly; they have been made as the minimum price for responsible speculation. But I want to ask in conclusion, why do almost all their speculations now point in the same direction? I have two answers, one cynical and one sentimental.

The cynic will say it is not accidental that the belief is held most strongly at a time when it is at last within the range of technology to discover if it is true. One of the reasons for our space program is to find out if there is extraterrestrial life; but unless the estimated probabilities of success are high, they cannot be used in justifying the enormous costs involved. What is more natural then, than that those who have to justify the costs or go out of the business of exobiology should strongly believe in the possibility, nay the probability, of success? I am not accusing anyone of dishonesty or even of disingenuousness; I am merely reminding you that nothing succeeds without hope of success, and that in general *the antecedent credulity with which an interesting hypothesis is held varies directly with the costs incurred in establishing it.*

The other answer is more speculative. While the old argument from design is heard no more, I suspect that deep-seated philosophical, religious, and existential commitments which once availed themselves of the argument from design are still silently effective in guiding the Lucretian argument and keeping alive the archetypal idea that man is not alone.[47]

Myth, religion, and now science-fiction with their tales of benevolent and malevolent extraterrestrial beings are commentaries on the human condition. I believe even responsible scientific speculation and expensive technology of space exploration in search for other life are the peculiarly modern equivalent of angelology and Utopia or of demonology and apocalypse.

In the sixteenth and seventeenth centuries there was a deep pessimism about the decline of nature, polluted by the sins of man.[48] Nature was redeemed if there were higher beings in the universe, so all was not lost even though man and earth were corrupt. There was the silence of infinite space which frightened Pascal; he suffered a 'Brunonian shock'[49] upon moving out of a friendly sphere into a lifeless infinite mechanism. This shock could be ameliorated by seeing the stars as other homes, and the universe as friendly to life after all.

We are now suffering from technological shock, destroying by radiological and chemical, if not moral, pollution the only abode of life we know. Are we not enough like our ancestors to respond with the same desperate hope they did? *Exobiology recapitulates eschatology*. The eschatological hope of help from heaven revives when the heavens of modern astronomy replace the Heaven of religion. That we can learn from more advanced societies in the skies the secret of survival is the eschatological hope which motivates, or at least is used to justify, the work of exobiologists.[50] But somewhat like people who object to spending money needed in the ghettoes on exploring the moon, I think the best hope for our survival is to be based on understanding human predicaments here on earth, not expecting a saving message from super-human beings in the skies.

Thinking about and even hoping to find extraterrestrial civilizations, however, sharpen our search for and appreciation of the peculiar virtues and vices of the only form of life we know. Exobiology and other exo-sciences cannot proceed merely by generalization from terrestrial experience; they must construct models of a more abstract nature of which terrestrial life and society are specifications. In that way hypotheses about extraterrestrial situations may throw light on the terrestrial, while the illumination of the extraterrestrial by hard facts about life on earth is at best dim and wavering. What Peter Winch has said about anthropology, we may say about exo-sociology: 'Seriously to study another way of life is necessarily to seek to extend our own – not simply to bring the other way within the already existing boundaries of our own . . . .'[51] Even if the exo-sciences fail to attain their prime goal, here is a valuable by-product. The quest for other, and better, forms of life, society, technology, ethics, and law may not reveal that they are actual elsewhere; but it may in the long run help us to make some of them actual on earth.

Yet after all there is some glimmer of hope for an answer. As long as it exists – and I think it will exist as long as we do – it would be a mistake to let niggardliness, skepticism, and despair inhibit the search. Many more harmful things can be done with our technology than listening for another civilization. If it should be successful, probably nothing is more worth using it for. So we have to ask, how should we proceed, and what shall we do if we succeed?

To the first, there are two simple and prudent answers. Let us give more thought to possible worlds so as to prepare ourselves to interpret any evidence we get that they are actual. Here is work for disciplined science-fiction writers, astronomers, biologists, psychologists, sociologists, and linguisticians. I venture to believe that even philosophers might be of some help.

Second, let there be world-wide sharing of resources of radio observatories. If all appropriate observatories devote some time to a systematic project of this kind, the costs in other more efficient research can be equitably spread.[52] But it must be remembered that the search is not worth undertaking unless it is planned to last decades, centuries, or even forever. Such cooperation would be a small step in bringing about the discovery of how much enlightened intelligence there is on one planet at least, our own. And what if we succeed? I have two conjectures. First, after a few weeks *it will be forgotten*, just as the details of the first moon landing have already been forgotten by most people. We are so well prepared by popular science and science-fiction for signals from outer space[53] that success will be just another nine-days' wonder like Orson Wells' 'Invasion from Mars' or the 'Great Moon Hoax' which shocked New York City in 1837.[54]

My second conjecture is: *it will never be forgotten*. For what is important is not a single discovery, but the beginning of an endless series of discoveries which will change everything in unforeseeable ways. We will be well prepared for the initial discovery, since we have to know what it will be to know when it has occurred. We are not prepared for the next discovery and the discovery after that; we have no idea what they will be. But if they are made, there is no limit to what in coming centuries we might learn about other creatures and, more portentously, about ourselves. Compared to such advances in knowledge, the Copernican and Darwinian Revolutions and the discovery of the New World[55] would have been but minor preludes.

Notes

1. *De generatione animalium* 761 b; *De motu animalium* 699 b 19; see also *Metaphysics* 1074 b 1–14.
2. *Syntagma philosophicum*, Part II, Sect. 2, Book 1, ch. 6; in *Opera omnia* (1658) vol. I, 524–530.
3. *Elements of Natural Philosophy*, ch. 3, end.
4. *Cosmologische Briefe* (1761), esp. letters 6, 8, and 9 (pp. 62–65, 93, 103, 113, 119).
5. *Allgemeine Naturgeschichte und Theorie des Himmels* (1755), Part III (not in English translation).
6. *Astronomy and General Physics Considered with Reference to Natural Theology* (Bridgewater Treatise, 1833), pp. 206, 207, 214.
7. The most notable is Roland Puccetti, *Persons* (New York: Herder and Herder, 1969).
8. *A Review of Space Research* (National Academy of Sciences–National Research Council Publication 1079 [1962]), ch. 9, pp. 2–3.
9. For example, Philolaos (Diels-Kranz [1964], p. 404); Anaxagoras (Kirk and Raven, p. 389); Democritus (*ibid.*, p. 410); Epicurus (Diogenes Laertius, ix, 31); perhaps Anaximander (see Charles Mugler, *Devenir cyclique et pluralité des mondes* [Paris, 1953]); *Timaeus* 41 E (see also *Laws* 967).
10. *City of God*, XII, ch. 11, 12.
11. Grant McColley, 'The Seventeenth-Century Doctrine of a Plurality of Worlds,' *Annals of Science* 1 (1936), 385–430 at 395.

12. *Summa theologica* I, Quest. 47, art. 3; see also Quest. 70, art. 3.
13. Bruno, *Of the Infinite Universe and the Worlds*, Fifth Dialogue (Singer transl.), p. 376; similarly, Nicholas of Cusa, *Of Learned Ignorance*, II, 13.
14. Good historical accounts will be found in McColley, *op. cit.*; A. O. Lovejoy, *The Great Chain of Being*, ch. 4; Marjorie Hope Nicolson, *The World in the Moon* (Smith College Studies in Modern Languages, XVII [1936]) and *Voyages to the Moon* (New York, 1948); and R. V. Chamberlin, *Life on Other Worlds, A Study in the History of Opinion* (*Bulletin of the University of Utah* 22, no. 3 [1936]).
15. *The Discovery of a New World, or A Discourse Tending to Prove (It is Probable) There May be Another Inhabitable World in the Moon* (1638) in *Mathematical and Philosophical Works of the Rt. Rev. John Wilkins* (1802, reprint, London, 1970).
16. *The Celestial Worlds Discover'd* (1698).
17. *Entretiens sur la pluralité des mondes* (1686).
18. *Essays*, Book II, xii (London, 1891), p. 226.
19. *Paradise Lost* VIII, lines 100–105.
20. *An Essay on Man*, Epistle I, lines 21–22 *et passim*.
21. *Civitas soli* (1623).
22. *Gulliver's Travels*, Book III: 'Voyage to Laputa.'
23. *Micromégas* (1752).
24. See Lucian's *Icaromennipus or the Sky-Man* as well as his better known *A True History*.
25. *The Age of Reason* (Liberal Arts Press, 1948), p. 44.
26. *Discourses on the Christian Revelation Viewed in Connection with Modern Astronomy* (1818) in *Works* (New York, 1850), IV, pp. 362–414.
27. Cited in note 6, *supra*.
28. *The Plurality of Worlds*, pp. 161, 172, 186.
29. *Ibid.*, p. 234.
30. *Ibid.*, pp. 282–287. The same problem led both Augustine (*loc. cit.*) and Melanchthon (*Initia doctrinae physicae*, in *Corpus reformatorum* XIII, columns 220–221) to deny the plurality of worlds. More imaginative solutions have appealed to some modern theologians; see the survey of (mostly German) literature in Wolfgang Müller, *Man Among the Stars* (New York, 1957), ch. 13.
31. For example, by Sir David Brewster, *More Worlds than One* (1854) and Richard Anthony Proctor, *Other Worlds than Ours* (1870) and *Our Place Among the Infinites* (1875).
32. *An Enquiry Concerning Human Understanding*, sect. 11.
33. Wallace's *Man's Place in Nature* (1893) is important not only for its conclusions against the doctrine of plurality, but because of its Lucretian rejection of teleological considerations which had been the principal bulwark of the theory before that time. It is the first treatment of the problem that reads like modern science. See J. M. Drachman, *Studies in the Literature of Natural Science* (New York, 1930), ch. 23.
34. See S. A. Kaplan, ed., *Extraterrestrial Civilizations* (Israel Program for Scientific Translations, Jerusalem, 1971), p. 257 n. (A very sophisticated book.)
35. See A. G. W. Cameron, ed., *Interstellar Communication* (New York: Benjamin, 1963); I. S. Shklovskii and Carl Sagan, *Intelligent Life in the Universe* (New York: Delta, 1964), ch. 27–35; S. A. Kaplan, *op. cit.*, pp. 1–212; Walter Sullivan, *We Are Not Alone* (New York: McGraw Hill, 1964), ch. 13–15. I cannot take seriously the possibility of establishing the existence of extraterrestrial civilizations by the observation of artifacts other than signals (e.g. 'Dyson spheres') because it seems to me we would be irresistably tempted by Occam's Razor to explain them as natural products. Only if we had 'direct' evidence (through intelligible signals) of the existence of extraterrestrial civilizations would an artificial origin of *other* artifacts appear to be a plausibly simple explanation. But I grant that the comparative simplicity of two hypotheses like these is an inexact notion, and one of them may appear more plausible at one time and the other at another. The fate of the 'canals' on Mars, however, does not inspire confidence in gross artifacts as evidence of intelligent design.
36. On alternative biochemistries see V. A. Firsoff, *Life Beyond the Earth* (London:

Hutchinson, 1963), pp. 106–146. On 'silicon life' see the remarks by Bergson in *Creative Evolution* (New York, 1911), pp. 256–257.

37. Compare the contrasting views of George Gaylord Simpson, 'The Non-Prevalence of Humanoids,' *Science* **143** (1964), pp. 769–775, and A. E. Slater, 'The Probability of Intelligent Life Evolving on a Planet,' *Proceedings of the VIIth International Astronautical Congress* (Barcelona, 1957), pp. 395–402, with R. Bieri, 'Humanoids on Other Planets,' *American Scientist* **52** (1964), pp. 425–458.

38. 'At the same time' is ambiguous. More precisely, the receiving civilization must at the time of reception be in the same technological stage as the sending civilization at the time of sending. Since stars are not all of the same age, the logic of the argument is not affected by the elapsed time of signal transport.

39. Compare the independent calculations of Shklovskii and Sagan (*op. cit.*, pp. 413, 418, 450) with those of Sebastian von Hoerner in Cameron (*op. cit.*, p. 275).

40. This denies that 'interstellar eavesdropping' (see J. A. Webb, in Cameron, *op. cit.*, ch. 18) will give us the desired evidence. The detection of signals meant for domestic consumption on a heavenly body is more probable than the detection of unidirectional signals beamed to us; but since the former cannot, in all probability, be deciphered, the evidence that they are intelligently modulated will be unavailable.

41. Hans Freudenthal, *Lincos: Design of a Language for Cosmic Intercourse* (Amsterdam, 1960), vol. I (all published); Lancelot Hogben, 'Astroglossa' in *Science in Authority* (London, 1963).

42. *Word and Object*, p. 26 '. . . it is to [nonverbal] stimulation that we must look for whatever empirical content there may be' in a language.

43. *Ibid.*, p. 52.

44. Suggested by Everett Hafner, 'Techniques of Interstellar Communication,' in *Exobiology* (Astronautical Society Publication no. 19 [1969]), pp. 37–68 at 60. Similar ideas in Hogben, *op. cit.*

45. Suggested by Philip Morrison in Cameron, *op cit.*, pp. 266–270; Sullivan, *op cit.*, ch. 18. The idea was developed by Fred Hoyle in his novel, *The Black Cloud* (New York: Harper, 1957).

46. Underlying the applicability of television is an anthropomorphic assumption that the information input of the sender of the message is like that of us human receivers, and there is not the slightest reason for believing this to be the case. Suppose, on the contrary, that his principal information channel is a sense organ unknown to us; or, for purpose of simplification, suppose it to be olfactory. In the latter case he identifies an apple by the distribution, spatial or temporal, of smells. He will then send a tele-olfaction image which we will interpret as a television image. But since the olfactory image he has is different from the optical image we receive, we will not recognize the latter as the image of an apple. (This assumes, furthermore, that there are things like apples up there.)

47. For somewhat similar ideas see Carl G. Jung, *Flying Saucers* (New York, 1969) and Robert Plank, *The Emotional Significance of Imaginary Beings* (Charles C. Thomas Co., 1968), esp. pp. 149–150.

48. See Victor Harris, *All Coherence Gone* (Chicago, 1949), esp. pp. 95, 104; Marjorie Hope Nicolson, *The Breaking of the Circle* (Evanston, 1950), ch. 3 and 4.

49. Wolfgang Philip, *Das Zeitalter der Aufklärung* (Bremen, 1963), p. xxvii.

50. For example, Gösta Ehrensvärd, *Man on Another World* (University of Chicago Press, 1965), pp. 7–9, 168 *et passim*; Puccetti, *op cit.*, pp. 113ff.

51. 'Understanding a Primitive Society,' *American Philosophical Quarterly* I (1964), 307–324 at 317–318.

52. Progress was made in this direction at the Byurakan Soviet–American Conference on Communication with Extraterrestrial Intelligence; see *News Report* (National Academy of Sciences) October, 1971, pp. 1, 4–5.

53. According to the Gallup Poll, 53 % of the people listed in *International Who's Who* believe in the existence of extraterrestrial life. *New York Times*, June 13, 1971.

54. Richard Adams Locke, *The Great Moon Hoax, or a Discovery that the Moon has a Vast Population of Human Beings* (New York, 1859).
55. Responses to the discovery of America, which was unanticipated, give, in spite of this difference, some clues to what may be reactions to the discovery of other worlds. For example, the question as to whether the American aborigines had souls and whether they were under natural law was discussed in terms not wholly unlike those in which our ethical and political relations with extraterrestrial beings are now discussed. See J. H. Elliott, *The Old World and the New, 1492–1650* (Cambridge University Press, 1970), especially chapter ii.

PART 2

Existence and nature of extraterrestrial intelligence

For there to be extraterrestrial intelligent life there must first be extra-terrestrial life; that is, it must be possible for life of some type to have originated elsewhere than on earth. I say 'of some type' to cover the possibility of life that is not carbon-based, as is all earthly life.

There is no generally accepted definition of life, nor is there agreement regarding on what a correct definition ought to be based – whether, for example, upon physiology, metabolism, biochemistry, genetics, thermo-dynamics, or indeed something else. Neither is there agreement on whether non-carbon-based life is possible. Carbon-based life is the only kind of which we know, but silicon life has been suggested as a possible alternative, and with the advent of computers – 'artificial intelligence' rooted in silicon chips – this suggestion may be thought to have some plausibility. Nevertheless, many scientists contend that because of the abundance of carbon in the universe, and its ease of bonding with other elements thereby forming a variety of stable compounds, carbon is the element of choice for the origin of life. Another problem with life based on alternative biochemistries is that we might not be able to recognize, much less interact with, such life even should it exist. For all these reasons, the authors in this section focus primarily on carbon-based life forms, 'life as we know it.'

How likely is it that life as we know it has arisen on other worlds? And if it has arisen elsewhere, what would such life be like? Asking these questions presupposes that there *are* other worlds – planets outside the solar system – for life to arise on. While no extrasolar planets are yet known to exist, theories of stellar formation have led astronomers to believe that many other stars have planetary systems. Given the abun-dance of stars in the universe, the chances are overwhelming that there are millions of potentially life-bearing planets. But does life exist on them?

For a while the main scientific obstacle to life on other worlds was the so-called 'combinatorial problem,' the problem of how – given the enormous number of possible combinations – the components of protein molecules could come together to fashion the proteins of living organisms. But at least in the case of earthly life they did so, and what happened on earth could certainly happen elsewhere. Lately, the work of Manfred Eigen and others has provided a theoretical explanation as to how this is possible.

Supposing then that life may arise on other worlds, what are the chances of its evolving into intelligent beings with whom we might communicate? Just over 20 years ago, Harvard biologist George Gaylord Simpson said that the chances are almost nil. In what many still consider to be the definitive presentation of this skeptical viewpoint, Simpson's paper 'The nonprevalence of humanoids' (*Science* [1964] **143**, 769) claimed that the random and highly idiosyncratic course of evolution on earth makes it wholly improbable that the evolution of intelligence could be repeated even here, much less on other planets where conditions would be entirely different. 'If human origins were indeed inevitable under the precise conditions of our actual history,' he wrote, 'that makes the more nearly impossible such an occurrence anywhere else.' Not only did Simpson say that 'duplicate men' were impossible elsewhere, he denied that any comparably intelligent organism would be likely to arise on other worlds.

In part two of this volume authors Ernst Mayr, David Raup, and Michael Ruse discuss the chances of intelligent beings evolving elsewhere. All three are specialists in evolution. Mayr, currently Emeritus Professor of Zoology at Harvard, changed biology's concept of what makes a species in his early work *Systematics and the Origin of Species* (New York: Columbia University Press, 1942; reprinted New York: Dover, 1964); his recent book, *The Growth of Biological Thought* (Cambridge: Harvard University Press, 1982), has been hailed as a masterpiece in the history of biology.

In his paper 'The probability of extraterrestrial intelligent life,' Mayr notes that, while it took intelligence millions of years to evolve here on earth, it evolved only once out of a billion species of earthly animals. But, if the existence of intelligence is an extraordinarily improbable phenome-non even on earth where the conditions for it are right, it must be even more improbable on those other worlds where the necessary conditions might not be so favorable.

Responding to those who argue that convergent evolution – of eyes, for example – supports the probable convergent evolution of intelligence, Mayr neatly turns the tables. The fact that eyes have evolved separately many times on earth, whereas intelligence has evolved only once, shows that the evolution of eyes is probable wherever they are of selective advantage whereas the evolution of intelligence – despite its being of enormous adaptive value – clearly is not.

Evolutionary paleobiologist David M. Raup, Chairman of the Department of Geophysical Sciences at the University of Chicago and coauthor with S. M. Stanley of the standard text *Principles of Paleontology* (San Francisco: W. H. Freeman, 2nd edn 1978), has a different interpretation of the course of evolution on earth. To begin, he suggests that the fossil record indicates that life itself may have originated independently several times on earth. This and other data make life on extraterrestrial bodies far more likely than had previously been thought. In answer to Mayr, Raup proposes a view of evolutionary convergence – of the sabertooth tiger and other animals – that *favors* the development of intelligence elsewhere.

Most striking of all, however, is Raup's idea that, contrary to what both SETI skeptics and SETI advocates had thought all along, it may be possible for us to detect radio transmissions from extraterrestrial life even though 'they' are without the advanced technology we normally associate with electromagnetic radiation. Raup's ingenious essay 'ETI without intelligence' will be controversial in both camps.

Supposing that intelligent extraterrestrial life does exist, what form is it likely to take? Have we reason to believe it will know and manipulate its environment as we do? Will sex be likely to develop on other worlds? And how about morality? Will ETIs have ethical standards? As philosopher of science Michael Ruse asks, 'Is rape wrong on Andromeda?'

Author of *The Darwinian Revolution* (Chicago: University of Chicago Press, 1979), *Darwinism Defended* (Reading, Mass.: Addison-Wesley, 1982), and other books on evolutionary science, Ruse is Professor of History and Philosophy at Canada's University of Guelph. In a wide-ranging, genial, but scientifically well-grounded essay, Ruse speculates on what ETs are really like. He finds there is reason to believe that extraterrestrials may be more like ourselves than even some SETI optimists are willing to imagine. Evolution, sexuality, epistemology, even morality, Ruse suggests, if not universal from one planet to the next, may be more common than previous treatments of the matter may lead us to suppose.

The probability of extraterrestrial intelligent life

ERNST MAYR

A number of very different problems are often confused during dis-cussions of the SETI project: (1) the probability of the existence of 'life' elsewhere in the universe, (2) the probability of intelligent extraterrestrial life, and (3) the chances of being able to communicate with such life, if it should exist.

At the present time we have no positive evidence whatsoever that life exists elsewhere, and thus, of course, also of intelligent life. The prob-abilities of either can be guessed at only by highly indirect inferences.

Life in the universe

When the Mars missions were being prepared, the astronomer Donald Menzel and I had a $5 bet as to whether or not 'life as on earth' [as was our precise designation] would be discovered on Mars. The physical scientist Menzel said yes, the evolutionary biologist Mayr said no. Who was right is on record. By now it is quite evident that none of the other planets in this solar system is suitable for life.

One negative instance, of course, proves nothing. If all suns in the universe have planets (actually a rather dubious assumption), we would have hundreds of millions of planets. Surely, it is argued, some of these should have spawned life. And I agree, the probability for a multiple origin of a self-replicating nucleic acid–protein aggregate is indeed high.

It has been known for some time that smaller organic molecules, like amino-acids, purines, and pyrimidines, can arise spontaneously in the universe, and that such processes can be duplicated in the laboratory. Nevertheless, for a long time it seemed impossible to explain how the amino-acids (and peptides) could get together with nucleic acids to form truly replicating, i.e. living, macromolecules. Through the researches of

Eigen and his school (Küppers 1983) there seems to be no longer a difficulty of principle. What is particularly interesting is the important role played by natural selection, even during the pre-biotic phase. The probability of the repeated origin of macromolecular systems with an ability for information storage and replication can no longer be doubted.

What is still entirely uncertain is how often this has happened, where it has happened, and how much evolution might have occurred subsequent to the origin of such life. We who live on the earth do not fully appreciate what an inhospitable place most planets must be. To be able to support life they must be just the right distance from their sun, have the right temperature, a sufficient amount of water, a sufficient density to be able to hold an atmosphere, a protection against damaging ultraviolet radiation, and so forth. Furthermore, every planet changes in the course of its history, and the sequence of changes has to be just right. If, for instance, there were too much free oxygen at an early stage, it would destroy life. The total set of prerequisites for the origin and maintenance of life drastically reduces the number of planets that would have been suitable for the origin of life. There is, indeed, the probability that the combination and sequence of conditions that permitted the origin of life on earth was not duplicated on a single other planet in the universe. I do not make such a claim, and it would not be science if I did, since it would be impossible ever to refute it. However, measured by the possibility of refutation, the claims of the proponents of extraterrestrial life and intelligence are equally outside the bounds of science. The only thing we know for sure is that of the nine planets of the solar system the earth is the only one that has produced life. Let us assume, however, for the sake of the argument that life has originated on some of the supposedly hundreds of millions of planets in the universe. Since we do not know how many suns have planets, the mentioned figure might be a gross overestimation.

The existence of extraterrestrial intelligence

It is interesting and rather characteristic that almost all the promoters of the thesis of extraterrestrial intelligence are physical scientists. They are joined by a number of molecular and microbial biologists, and by a handful of romantic organismic biologists.

Why are those biologists, who have the greatest expertise on evolutionary probabilities, so almost unanimously skeptical of the probability of extraterrestrial intelligence? It seems to me that this is to a large extent due to the tendency of physical scientists to think deterministically, while organismic biologists know how opportunistic and unpredictable evolution is.

Some 20 years ago when I argued a great deal with the astronomer Donald Menzel about life on Mars, I was forever astonished how certain he was that if life had ever originated on Mars (or been transported to Mars), this would inevitably lead to intelligent humanoids. The production of man was for him like the end product of a chemical reaction chain where the end product can be predicted once you know with what chemicals you had started. He took it virtually for granted that if there was life on a planet it would in due time give rise to intelligent life: 'Our own Milky Way might contain up to a million planets [favorable to the development of life], all inhabited by intelligent life' (Menzel 1965, p. 218).

Everybody knows, of course, that determinism is no longer the fashion in modern physics, and yet in conversation with physical scientists I have discovered again and again how strongly they still think along deterministic lines. If organic evolution on earth culminated in intelligence, why should it not have resulted in intelligence on all planets on which life had originated?

By contrast an evolutionist is impressed by the incredible improbability of intelligent life ever to have evolved, even on earth. To demonstrate this, let us look at the history of life on earth.

Date of origin of kinds of organisms if age of earth (4.5 billion years) is made equivalent to a calendar year:

> Origin of
> Earth = 1 January
> Life (Prokaryotes) = 27 February
> Eukaryotes = 28 October
> Chordates = 17 November
> Vertebrates = 21 November
> Mammals = 12 December
> Primates = 26 December
> Anthropoids = 30 December, at 01:00 a.m.
> Hominid line = 31 December 10:00 a.m.
> *Homo sapiens* = 31 December 11.56½ p.m. (= 3½ minutes
> before year's end).

Let us look at the chronology of major evolutionary events on earth. (All cited figures are rough estimates, the upward or downward revision of which would have no effect on the argument; the order of magnitude, however, is right.) Let us assume the earth originated 4.5 billion years ago. There is evidence that life began only about 700 million years (*my*) later. Definite early prokaryote fossils are known from 3.5 billion years ago. What is most remarkable is that for about 3000 *my* nothing very spectacular happened as far as life on earth was concerned. There was apparently a rich diversification of prokaryotes, but these – although quite successful in their way – are poor potential as progenitors of intelligent life. Nevertheless, they displayed remarkable metabolic diversification, the blue-green bacteria even became phototropic and produced oxygen. Up to that time the earth's atmosphere had been reducing.

Sometime, between 800 and 1000 *my* ago a most improbable event took place. According to the most likely explanation, a symbiosis was established between two (or more) kinds of prokaryotes, one of them supplying cytoplasmic organelles, the other one the nucleus of an entirely new type of organism, the first eukaryote. This was apparently such a successful combination that within a period of about 100 *my* (estimate) four new kingdoms evolved, the protists (one-celled animals and plants), fungi, plants, and animals. All higher organisms are eukaryotes, characterized by the possession of a well-organized nucleus and chromosomes in each cell.

We can see that from the origin of life to the origin of the eukaryotes about two-thirds of the age of the earth had passed by without any noticeable events except for diversification within the prokaryotes. But once the eukaryotes had been 'invented' an almost explosive innovative diversification took place. Within each of the four mentioned kingdoms scores of separate evolutionary lines originated, many of them strikingly different from each other. However, in none of these kingdoms, except that of the animals, was there even the beginning of any evolutionary trends toward intelligence.

What about evolution of intelligence among the animals? After the animalian 'type' had been invented, different structural types originated with such fertility that one could probably recognize at least 40 different phyla of animals in the Cambrian, including unique types in the Ediacaran and surviving Burgess shale formations. Many of these became extinct rather quickly and there is no good evidence for the origin of any new phylum after the end of the Cambrian (500 *my* ago). However, the surviving phyla experienced a continuing abundant proliferation into classes, orders, families, and lower taxa. Of the 40 or so original phyla of animals only one, that of the chordates, eventually gave rise to intelligent life, but the world still had to wait some 500 *my* before this happened. At first, still in Paleozoic, the vertebrates appeared in exceedingly diverse types, formerly all lumped together under the name 'Fishes', but it is now realized how different the early vertebrates were from each other. Among this multitude of types only one gave rise to the amphibians, and among the various types of amphibians only one to the reptiles. What are called the reptiles are again a highly diverse group of vertebrates including such different organisms as turtles, lizards, snakes, crocodiles, and numerous extinct lineages as ichthyosaurs, plesiosaurs, pterodactyls, and dinosaurs. Among these numerous types of reptiles, only two, the pseudosuchians (ancestors of birds) and the therapsids (ancestors of mammals) gave rise to descendants to some of whom a reasonable degree of intelligence can be attributed. But with all my bias in favor of birds, I would not say that a raven or parrot has the amount and kind of intelligence to found a civilization. So we have to continue with the mammalian class. It contains such unusual types as the monotremes (e.g. platypus) and marsupials, as well as a rich assortment of placental orders, some still living, many others having become extinct in the course of the Tertiary. Forms with a rather high development of the central nervous system and a good deal of intelligence are quite common among the mammals, but only one of these many orders led to the development of a truly superior intelligent life, the primates. The primates, however, are a rather diversified group, with prosimians (lemurs, etc.), New World monkeys, and Old World monkeys, but only the anthropoid apes produced intelligence that clearly surpasses other mammals. Only after 18 of the 25 *my* of the existence of the anthropoid apes, and after a splitting of this major lineage into a number of minor lineages, like the gibbons (and relatives), the orang utan (and relatives), the African apes (chimpanzee and gorilla), and a considerable number of extinct lineages, did the lineage emerge which eventually, less than one-third of a million years ago, led to *Homo sapiens*.

The reason why I have buried you under this mass of tedious detail is to make one point, but an all-important one. In conflict with the thinking of those who see a straight line from the origin of life to intelligent man, I have shown that at each level of this pathway there were scores, if not hundreds, of branching points and separately evolving phyletic lines, with only a single one in each case forming the ancestral lineage that ultimately gave rise to Man.

If evolutionists have learned anything from a detailed analysis of evolution, it is the lesson that the origin of new taxa is largely a chance event. Ninety-nine of 100 newly arising species probably become extinct without giving rise to descendant taxa. And the characteristic of any new taxon is to a large extent determined by such chance factors as the genetic composition of the founding population, the special internal structure of its genotype, and the physical as well as biotic environment that supplies the selection forces of the new species population.

My argument based on the incredibly low probability of life ever having originated on earth, must not be misunderstood. I do not claim in the least that an extraterrestrial intelligent 'life' must have the slightest anatomical similarity to man. I already mentioned that we get a certain amount of intelligence in other mammals and even in birds. Indeed, an ability to make use of previous experience in subsequent actions, in other words a rudimentary kind of intelligence, is widely distributed in the animal kingdom. Intelligence, on another planet, might reside in a being inconceivably different from any living being on earth. Any devotee of science fiction will have no trouble in coming up with possibilities.

The point I am making is the incredible improbability of genuine intelligence emerging. There were probably more than a billion species of animals on earth, belonging to many millions of separate phyletic lines, all living on this planet earth which is hospitable to intelligence, and yet only a single one of them succeeded in producing intelligence.

Proponents of extraterrestrial intelligence have mentioned the convergent evolution of two such 'highly improbable' organs as the eye of the cephalopods and of the vertebrates as an analog to the presumably equally improbable but not at all impossible convergent evolution of intelligence. Those who have thus argued, unfortunately, do not know their biology. The case of the convergent evolution of eyes is, indeed, of decisive importance for the estimation of the probability of convergent evolution of intelligence. The crucial point is that the convergent evolution of eyes in different phyletic lines of animals is not at all improbable. In fact, eyes evolved whenever they were of selective advantage in the animal kingdom. As Salvini-Plawen and I have shown, eyes evolved independently no less than at least 40 times in different groups of animals (Salvini-Plawen & Mayr 1977). This shows that a highly complicated organ can evolve repeatedly and convergently when advantageous, provided such evolution is at all probable. For genuine intelligence this is evidently not the case, as the history of life on earth has shown.

One additional improbability must be mentioned. Somehow, the supporters of SETI naively assume that 'intelligence' means developing a technology capable of intragalactic or even intergalactic communication. But such a development is highly improbable. For instance, Neanderthal Man, living 100000 years ago, had a brain as big as ours. Yet, his 'civilization' was utterly rudimentary. The wonderful civilizations of the Greeks, the Chinese, the Mayas or the Renaissance, although they were created by people who were for all intents and purposes physically identical with us, never developed such a technology, and neither did we until a few years ago. The assumption that any intelligent extraterrestrial life must have the technology and mode of thinking of late twentieth-century Man is unbelievably naive.

Civilizations, as human history demonstrates, are fleeting moments in the history of an intelligent species. For two civilizations to communicate with each other, it is necessary that they flourish simultaneously. Let me illustrate the importance of this point by a little fable. Let me assume there is another high technology civilization in our galaxy. By some extraordinary instrumentation their inhabitants were able to discover the origin of the earth 4.5 billion years ago. At once they began to send signals to the earth and continued to do so for 4.5 billion years. Finally at the time of the birth of Christ they decided that they would terminate their program after another 1900 years, if they had not received any answer by then. When they abandoned their program in the year 1900, they had proven to their own satisfaction that there was no other intelligent life in our galaxy.

I am trying to demonstrate by this fable that even if there were intelligent extraterrestrial life, and even if it had developed a highly sophisticated technology [although if they were truly intelligent they would probably carefully avoid this], the timing of their efforts and those of our engineers would have to coincide to an altogether improbable degree, considering the amounts of astronomical time available. Every aspect of 'extraterrestrial intelligence' that we consider confronts us with astronomically low probabilities. If one multiplies these with each other, one comes out so close to zero, that it is zero for all practical purposes. This was already pointed out by Simpson in 1964. Those biologists who doubt the probability of ever establishing contact with extraterrestrial intelligent life if it should exist, do not 'deny categorically the possibility of extraterrestrial intelligence', as they have been accused. How could they? There are no facts that would permit such a categorical denial. Nor have I seen a published statement of such a 'categorical denial'. All they claim is that the probabilities are close to zero. This is why evolutionary biologists, as a group, are so skeptical of the existence of extraterrestrial intelligence, and even more so of any possibility of communicating with it, if it exists.

For all these reasons I conclude that the SETI program is a deplorable waste of taxpayers' money, money that could be spent far more usefully for other purposes.[1]

Note

1. After completing the manuscript, I reread Simpson's classical paper (1964) on the subject and was struck by how similar his analysis was to mine. Perhaps subconsciously I still remembered his arguments. Be that as it may, I warmly recommend reading his famous essay on the 'Nonprevalence of humanoids'. It is as pertinent today as it was 20 years ago.

30   *Ernst Mayr*

References

Küppers, B.-O. (1983). *Molecular Theory of Evolution*. Berlin, Heidelberg, New
   York: Springer.
Menzel, D. H. (1965). Life in the Universe. *The Graduate Journal*, 7, 195–219.
Salvini-Plawen, L. v. & Mayr, E. (1977). On the evolution of photoreceptors and
   eyes. *Evolutionary Biology*, **10**, 207–63.
Simpson, G. G. (1964). The nonprevalence of humanoids. In *This View of Life*,
   pp. 253–71. New York: Harcourt, Brace, and World.

ETI without intelligence

DAVID M. RAUP

'Any competent evolutionary biologist should acknowledge the overwhelming improbability of humanoids elsewhere in space.'

I made up this quote, but close counterparts can be found throughout the literature of the SETI debate. Evolutionary biologists play a curious role in any discussion of extraterrestrial intelligence. Most are quite negative and their views are often trotted out by the anti-SETI forces to argue for the futility of search programs, or at least for the very small probability of success. Some of the truly great thinkers in evolutionary biology, including Simpson, Mayr, Dobzhansky, and Ayala have felt that there have been too many unpredictable events in the evolution of the humanoid condition for that process to be repeated in a different and independent biological system. The significance of this depends, however, on exactly what is meant by 'humanoid,' and this has led to some confusion and misunderstanding. The main argument made by Simpson and others is that the independent evolution of another organism having the *physical form* of *Homo sapiens* is most unlikely. To have this combined with humanoid *intelligence* is seen as even less likely. But what is not usually understood is that the SETI strategy calls only for the *intelligence* part of the humanoid condition: the *physical form* of the alien could be totally unlike *Homo sapiens*. Evolutionary biologists have not addressed the intelligence issue by itself in any detail and thus one cannot say categorically that all (or most) evolutionary biologists are opposed to the SETI concept.

In this essay, I will consider several aspects of the problem of repeated evolution of the humanoid condition, whether humanoid be defined in terms of physical form or intelligence or both. But my main purpose will be to present some ideas about a different and distinctly non-human kind of intelligence. If such intelligence exists, we may have a good chance of detecting it even though it represents a departure from the normal SETI strategy.

The evolution of complex life on earth

There now seems to be a reasonable consensus among biologists and biochemists that the spontaneous origin of life is fairly straightforward chemically and that hence primitive life, at least, in other planetary systems is possible and even quite probable. There are several arguments favoring this point of view, including the discovery of organic compounds in meteorites, but one of the strongest comes from the fact that recognizable fossils are found in almost the oldest well-dated rocks on the earth. Because life originated very early in the history of the earth, its beginning may not have been difficult. Let me explain.

The earth is now estimated to be about 4.5 billion years old, but we have not been able to date rocks older than about 3.8 billion. This discrepancy is not serious and stems largely from the fact that as we go back in time, there are fewer and fewer rocks that have survived the ravages of erosion and metamorphism well enough for their mineral assemblages to retain an accurate isotopic record of their origin. It is easy to find an early record of something but often nearly impossible to find the *first* occurrence of something. The oldest fossils are in rocks about 3.5 billion years old, almost as old as the oldest dated rocks. Because the chances of finding recognizably preserved fossils in rocks this old are also very small, we cannot attach much significance to the fact that the oldest fossils are somewhat younger than the oldest rocks. It may even be that life originated considerably earlier. The approximate coincidence of ages is thus strong support for the idea that life originated very shortly after the earth was formed and this in turn supports the contention that life origination is 'easy.' To make things even easier, a recently published set of calculations (Raup & Valentine, 1983) argues that life may have originated several times independently on the early earth and that, if so, we would not be aware of it now because all life today would have descended from just one of the origination events.

The foregoing is speculative, of course, but I will work from the assumption that life should be a fairly common phenomenon in other solar systems, given the proper physical and chemical conditions. This last point is critical in estimating the probability of life on other planets. My earlier comments refer to the likelihood of life *like ours* originating and surviving. This means a carbon-based system, self-replicating nucleic acid–proteins aggregates, and so on. It also implies physical and chemical conditions suitable for life as we know it, including restricted ranges in temperature, pH, and other parameters. But, I submit, we have to be very careful talking about this for two reasons. First, we may not have a full grasp of the range of conditions under which our kind of life can exist and, secondly, we cannot ignore the possibility that some sorts of life may operate on a different physical and chemical basis and thus have totally different requirements and limitations.

On the first point, biologists have been somewhat shaken in recent years to discover organisms living apparently quite happily in some very extreme environments, such as hot springs and deep-sea hydrothermal vents. With each such discovery, the range of conditions suitable for life has had to be expanded. On the second point, it is a difficult task to try to imagine and analyze life systems totally different from our own. Is complex life without DNA possible? Must the basic building blocks be chemically similar? Suffice it to say that there is a mass of uncertainty on these questions. When we talk about the probability of life elsewhere in space, we tend to limit our thinking to our kind of life. If we grant that other kinds of life could exist and that these other kinds might be able to develop complex forms capable of some sort of intelligent action, then our current estimate of the probability of life – any life – is almost certainly on the low side.

We are in a slightly better position to grapple with the problem of evolving complex life once we have life in the first place. For at least one planet (the earth), we have an excellent fossil record of that evolution. Critical questions are: Need it have taken three billion years to have developed complex, multicellular organisms, given the primitive forms of the oldest fossils? Need it have taken an additional half billion years for intelligence to have evolved? Are there peculiarities of this planet and its environmental history that have influenced the tempo and mode of evolution? Definitive answers to these and other questions are elusive, of course, but we can get somewhere.

For almost 3 billion years after life originated, it was a pretty dull operation, at least to me (though not to the biochemist or microbiologist). It was not until the late Precambrian, perhaps 600–700 million years ago, that fossils of complex multicellular animals appear in the rock record. By the early Cambrian period, about 550 million years ago, tremendous evolutionary strides had been made and the oceans were inhabited by a remarkably modern array of shellfish, sophisticated arthropods, and so on. Marine animals had evolved to a point where the casual lay observer might not see much difference from life now. To be sure, there were no land plants nor any fishes, birds, or other vertebrate animals. But complex biology was off and running.

Between the Cambrian and now, many changes in the earth's biota have taken place. From the human point of view, at least, the most important events were the evolution of land plants and the development of diverse vertebrate groups including, of course, *Homo sapiens*. One can see order and directionality only if the evolutionary record is viewed from a great distance. That is, if one considers a several billion year span, there are net increases in overall complexity and diversity. But when the record is viewed in more detail, there are strong elements of apparent randomness and even chaos. Plant and animal groups came and went, some spectacular inventions appeared and a few, at least, seem to have had profound effects on other elements of the biosphere. But there was anything but a neat and simple progression from simple to complex or from unsophisticated to sophisticated. Some innovations, like flying, appeared early and frequently, while others either never appeared (effective wheeled vehicles, aquatic animals with truly efficient wind powered sailing systems, etc.) or they appeared later in geologic time than seems necessary. If plant and animal evolution were to be replayed all over again, the results would certainly be different and probably strikingly so.

The general lack of orderliness is an important point. The evolution of any lineage, and even its very survival, is the result of the organism's reaction to a host of events and changes in the physical and biological environment. Even if the sequence of environmental changes is the same, the organism may react differently because of differences in genetic makeup or because of pure chance. All this makes the course of evolution basically unpredictable. It is rather like the history of human civilizations: some explanations of specific changes can be found, after the fact, but prediction of future change is nearly impossible, except in a very general statistical sense, because of the sheer complexity of causal factors.

To cite just one example, the mass extinction at the end of the Cretaceous period (65 million years ago) killed off many animal groups, including the dinosaurs. The mammals lost many species at the same time, so many that they might easily have gone completely extinct. We have no evidence that the mammals of the time were any better adapted than the dinosaurs, except for the fact that mammals did survive and dinosaurs did not. The outcome of the Cretaceous mass extinction may well have been a matter of pure chance or an accident of differing susceptibility to whatever special stresses caused the overall mass extinction. It is somewhat frightening to realize that if the end of the Cretaceous were to be replayed, our own mammalian ancestors might not have survived.

It is clear that evolution does not follow any set track. Nothing is inevitable or determined in advance. In particular, we cannot say that intelligence will necessarily evolve in other worlds even though life in those other worlds may have similar starting points and environmental histories. Even if intelligence does evolve in another world, we have no basis for saying how long this should take. Thus, the value of studying the evolutionary record on earth is limited primarily to a general assessment of the range of things that are possible in evolution.

Humanoids and evolutionary convergence

Intelligent creatures from outer space are often depicted as being surprisingly like *Homo sapiens*. They may be green or have strangely placed eyes or even extra arms, but they are still recognizably like earthly humans – or, at least, more similar to humans than to any other animals on earth. This general idea has been promoted heavily by science fiction writers but it also stems from the common misconception that any evolutionary system will follow a predictable sequence leading to something like *Homo sapiens*. This notion runs counter, of course, to the claim I made earlier that if evolution were to be replayed, the results would be strikingly different. And it is in this area that the leading evolutionary biologists have been so adamant in their criticisms of SETI-type scenarios. G. G. Simpson wrote (1964, p. 268), for example, that: '. . . any close approximation of *Homo sapiens* elsewhere in the accessible universe is effectively ruled out . . . .' Although I am in general agreement with this view, a caveat about the phenomenon of *evolutionary convergence* is necessary.

Even though the course of evolution is unpredictable, we do know that certain types or designs of organisms have evolved more than once. An example is the well-known sabertooth tiger. The common fossil sabertooth preserved in the La Brea tar pits of Los Angeles was a placental mammal and although it is extinct, we know a great deal about its anatomy. In South America at about the same time geologically, there was a marsupial version of the sabertooth tiger. Surprisingly similar anatomy evolved independently in the two mammalian groups. Although placental and marsupial mammals have a common ancestor in the Mesozoic era, they had been genetically separate for tens of millions of years before the sabertooth form appeared. This is what is known as evolutionary or adaptive convergence. In the sabertooth case, similar anatomy apparently evolved in response to similar environmental pressures and/or opportunities. Both evolutionary lines were faced with the same problems and both 'chose' approximately the same solution.

It is presumed that convergence is most common in situations where there are only a few ways of solving a particular problem, thus increasing the probability that independent lineages will adopt the same solution. One can make an analogy between this and the situation where isolated human cultures come up with the same inventions – except that in the human case the mechanism is cultural rather than biological evolution.

Evolutionary convergence is an important element of any consideration of a search for extraterrestrial life. The optimists claim that humanoid anatomy and intelligence should be expected to evolve repeatedly. They claim that if it has happened once, it could happen many times. This says, in effect, that the humanoid condition is one that should be subject to evolutionary convergence. As I noted earlier, the current American SETI program is banking on convergence in intelligence but not anatomy. The counterclaim of the pessimists is that the humanoid condition is a highly specialized one, the culmination of literally millions of inter-connected evolutionary choices, and therefore most unlikely to be repeated in our own biological system or any other. A strong point used to support this side of the argument is that nothing approaching the humanoid condition has evolved on earth except in the hominid line. This is only a statistical argument, of course, but it is a potent one in view of the large number of separate lineages that have evolved on earth.

Unfortunately for the general problem on humanoid convergence, we know surprisingly little about evolutionary convergence as a process. We have many anecdotal examples, such as the sabertooth tiger, but evol-utionary biology has not yet reached the point of being able to predict which anatomical configurations are likely to be repeated and which are not. We can only work with the statistical experience of the last 550 million years of evolution on earth. Therefore, any assessment of the likelihood of humanoids developing in other biological systems comes down to guesswork based on somewhat limited experience. At the moment, the majority of evolutionary biologists find the chance of an independent evolution of a recognizably humanoid creature to be essen-tially nil. But we must be cautious.

An interesting case of an unconventional view of humanoid con-vergence was developed a couple of years ago by Dale Russell of the Canadian National Museum. Dr Russell asked the question: If the large reptiles (dinosaurs) had survived the extinctions at the end of the Cre-taceous period, what would they look like now, 65 million years later? He realized full well that it is risky to try to answer such a question, but try he did. He produced a scale model of what he thought one particular dinosaur (*Stenonychosaurus*) might have become. The model attracted quite a bit of public attention because it looked remarkably humanoid: upright posture, a rather humanoid face, a large head, and so on. Was Russell just a victim of anthropomorphism or was he demonstrating, in effect, that the humanoid condition should be expected to evolve as an adaptive convergence? It should be kept in mind that Russell was working with the evolution of a group, reptiles, that is much more similar to mammals to begin with than would be the case if one were considering a totally alien biology. Even so, most evolutionary biologists do not agree with Russell's projection of evolution – or even grant the possibility that such projection is possible – but we have the proposal and it is made by a highly competent anatomist and paleontologist using the best information available (for detail, see Russell & Seguin 1982). In my view, we know so little about the phenomenon of convergence that we should be extremely cautious in making claims about what is possible and what is not. Above all, we should try to learn more about the evolutionary biology of the humanoid condition.

Technology without intelligence

The current American SETI program is rather modest in that it seeks only to *listen* for radio transmissions from distant planets. In fact, much of the present strategy is oriented toward picking up accidental leakage of radiation no more powerful than that which now leaks into space from our own radio, television, and radar transmissions.

In spite of the simplicity of the SETI plan, certain critical assumptions about alien biology have been deemed necessary. It has been assumed that, for SETI to be successful, species on other planets must have evolved the advanced civilizations and high intelligence necessary to produce a technology at least as sophisticated as our own. Because radio transmitters, radars, and radiotelescopes are artificial (non-biological) constructs, the alien species would have to have much of the elaborate tool-making ability of *Homo sapiens*, along with an intelligence capable of independent, conceptual thought and reasoning.

One of the important NASA analyses of the problem (Morrison *et al.*, 1977) put it this way (pp. 5–6), referring to the kind of radio signature SETI is looking for:

> '. . . such a radio beam cannot come, we think, from any glowing sphere of gas or drifting beam of particles. It can come only from something like our own complex artificial apparatus . . . something we would recognize as the product of other understanding and ingenious beings.'

In other words, the alien being is visualized as having a strikingly humanoid intelligence. [But note that the humanoid physical form is not set up as a requirement.] The proposition quoted above has produced all sorts of debate on the likelihood of such a biological system evolving elsewhere in space. It is again a question of evolutionary convergence. Honest and informed arguments have been presented on both sides but to me, at least, no truly compelling case has been put forward for either side.

Even granting the evolution of intelligence, there is the question of whether this intelligence would lead to a technology that we could detect from earth. A particularly optimistic statement on this is also contained in the NASA document cited above (*ibid.*, p. 51):

> 'Almost certainly once a species with the requisite intelligence, manipulative ability, and complex social organization has evolved, technological civilization will develop . . . thus, once a system capable of conceptualizing sophisticated internal models of external phenomena has evolved, it is only a matter of time before all possible ideas inherent in the available sensory perceptions are conceived.'

This is a tall order! And I have to agree with the evolutionary biologists who feel that this calls for a highly deterministic view of evolution. It could happen, but the probability is likely to be low.

To make matters worse, successful detection of radio or other transmissions requires that we, and the alien being, operate in the same geologic time frame. Much has been said and written on this question. Since there is nothing in the evolutionary process to guarantee that technology will develop at the same time in different systems, the focus of attention lies with the expected longevity of technological civilizations. The conventional wisdom is that our own technology will be short lived – either because of self-destruction or because the technology will change by cultural evolution into something quite different. In the latter case, it has been suggested that the present level of leakage of electromagnetic radiation into space may be temporary because of the development of tighter, more efficient means of electronic communication. Clearly, the problem of timing is critical.

Let me suggest that the intelligence aspect of the SETI reasoning may be tackled in a way quite different from the normal SETI manner. The essential point is that we may be confusing *what* we are looking for (radiowaves of biological origin) with *how* they are generated (humanoid intelligence).

It is widely assumed that man-like intelligence and advanced civilizations are essential for the production of detectable signals in space. Are there other ways? I think so, at least in theory. To explore this, let us look more carefully at intelligent behavior and focus not on *how* it works but on *what* it produces. Reduced to bare essentials, intelligent behavior produces *solutions to problems* which enable the individual or the species to survive or perform better. Problem-solving ability is a key part of most definitions of human intelligence. An intelligent species is able to react successfully to experience and to respond to new situations. We need not be concerned with how this is done and can thus avoid words like *reason*, *understanding*, *thought*, and *conceptualization*.

A large variety of organisms now living on earth exhibit behavior which mimics intelligent behavior even though these organisms are not intelligent in our terms. I am thinking of the stereotyped, hard-wired behavior seen most strikingly in the social insects but also present to some degree in almost all organisms. As we shall see, animals have solved many of the same problems that we have (and some that we have not), but they have done it by a different means: adaptation through natural selection. Let me give some examples.

Protective mimicry is a common phenomenon. A butterfly, for example, may achieve immunity from predators by evolving a color pattern which mimics the appearance of a poisonous species known and recognized by predators. The predator avoids all butterflies with that particular color pattern, regardless of species. Mimicry evolves over a long series of generations by selecting those chance mutations that make the non-poisonous species look more like the poisonous one. In the process, many butterflies are eaten by predators but the result is the enhanced survival of the species. Exactly the same result could have been achieved by an intelligent organism. If butterflies were intelligent in the human sense, they could have reasoned out the problem and perhaps arrived at the same solution, implementing it by genetic engineering. Alternatively, the hypothetical intelligent butterfly could have manufactured artificial camouflage (perhaps painting its wings) to accomplish the same thing. Note that there is no practical difference between manufacture of camouflage by a technologically inclined species and the growing of camouflaging coloration as part of the organism's metabolism. They are both manufacturing. The main point is that the problem of protection can be solved either by intelligence or by standard Darwinian adaptation.

Using intelligence to solve problems is much faster than using natural selection and we, as members of an intelligent species, set great store by our ability to react quickly to new situations. But when we consider the vastness of geological time, is the difference between minutes, hours, years, or even millions of years significant? Butterflies solve problems, as they arise, whether the problems involve the appearance of new predators or changes in climate. Some insect species have certainly become extinct because they could not react quickly enough, but many others have survived, and survived well.

To move on to other examples, many social insect species have achieved some truly remarkable feats of engineering. Bees and termites, operating in large, organized societies, have developed sophisticated systems of climate and air quality control in the hive or nest. Certain ant species maintain elaborate slave societies with complex systems of raids on other species to replenish the slave populations. The list of examples is endless and familiar to any watcher of television nature programs. In all these situations, the individual organisms are not intelligent. They are programmed by their genetic inheritance to do certain things at certain times or in response to specific stimuli but the results are not really different from what an intelligent organism might produce. To be sure, the structure of insect societies is repugnant to most of us – with an almost total lack of individual will – but this is not relevant in the present context.

Another class of examples is a little closer to the SETI problem. A number of animals 'manufacture' instruments for navigation and communication which are remarkably like systems made by human beings. The recently discovered use of magnetic materials by some birds, fish, and insects represents a surprisingly sophisticated technology. It is not a technology based on artificially fabricated instruments, as ours would be, but it accomplishes the same thing. These animals manufacture, as part of their physiology, the magnetic mineral magnetite and use it to measure the direction and intensity of the earth's magnetic field. More important, they have evolved the software necessary to interpret and use this information in long distance navigation. As J. L. Gould (1980) has put it, these animals: '. . . are most usefully thought of as no more than microcomputer equipped machines, programmed and then laboriously "debugged" by their genes . . . .'

Although the migratory birds and other animals developed navigational systems based on magnetite by the crude and slow method of natural selection, it may be worth pointing out that they solved the problem millions of years before we did! The process of adaptation through natural selection is a remarkably effective scheme for finding optimum solutions to engineering problems.

Many species have also evolved the ability to emit (or transmit) sound and various forms of electromagnetic radiation. Sonar systems are well known in bats, some marine mammals, and in other animals. Many fish, including sharks, rays, catfish, eels, and several families of the so-called 'electric fish' can sense weak electric fields. The electric fish generate electric fields themselves and use these fields for hunting and communication. Eels are capable of giving off electric impulses of as high as 600 volts.

The discoveries of most of the phenomena I have just described came as complete surprises to biologists. Before Lowenstam discovered the mineral magnetite in marine organisms, the biological synthesis of this mineral would have seemed incredible. A number of substances are regularly synthesized by organisms which are difficult or (so far) impossible to produce inorganically in the laboratory under earth-surface conditions. And the computer-like software that has evolved in many organisms is mind-boggling.

What does all this have to do with SETI? I think it shows that there is at least one alternative route to sophisticated technology in biological systems. It is the utterly simple, familiar process of adaptation and has no requirements of intelligence! Furthermore, the complex adaptations are found throughout the biological world and many represent solutions to what to us are common engineering problems. Some of the most elegant adaptations have evolved independently many times, including sonar and the use of electric fields.

So, I am suggesting that any evolving biological system has the potential of sending detectable evidence of its presence into space. You may argue that the sort of adaptive phenomenon I am describing would be unlikely to lead to the sending of radio or other signals into space. Why would an animal evolve an awareness of the space environment? What adaptational problem could be solved by producing the technology necessary to penetrate space? The answers to these questions are not straightforward but one aspect should be noted. Radio capability was developed by the human species solely for earthbound communication between members of the species. It was only later that the technology was converted to the investigation of space (radioastronomy). Also, most of the transmissions from the earth are inadvertent leakages from communications systems. So, an interest in space *per se* is unimportant and unnecessary.

Could animal life as we know it produce enough electromagnetic or other radiation to be detected from space? We do not know this but the history of biology is so full of surprises that it would be folly to rule out the possibility. Consider that the CB (Citizen's band) radio transmitter operated by a 12-volt battery in any taxicab has the potential of being detected in space, although transmission of such a signal from our planet is heavily screened by the earth's ionosphere. It may be that detectable signals from an alien species require more energy than could be produced by any of the animals we know on earth. But with the wide range of sizes possible for individual organisms, to say nothing of large aggregates (colonies), I do not think we can place limits on the potential of an alien species.

42   *David M. Raup*

Getting away from a dependence on searching for humanoid intelligence means that we can avoid the crucial problem of timing, discussed earlier. The kind of evolved animal technology I am suggesting could survive for long periods. It could exist and thrive for tens or hundreds of millions of years, as lesser versions have done on earth. And the problem of self-destruction is remote.

Furthermore, our earth is associated with a relatively young star. This means that many other planetary systems have had much more time for the kind of evolutionary development I am considering here. Although this added time provides no guarantee, it does enhance the probability of there being biological achievements beyond what we observe on earth.

In summary, I propose that the *manifestations* we ascribe to an intelligent being, and which are crucial to the SETI strategy, can be produced by an *un*intelligent organism and the mechanism for accomplishing this is the ubiquitous process of adaptation. For a search for extraterrestrial life to be successful, we do not have to assume an elaborate development through cultural evolution of the kind of contemplative, advanced society that we have on earth. Whereas this new perspective does not guarantee success, I think it substantially improves the chances for success.

References

Gould, J. L. (1980). The case for magnetic sensitivity in birds and bees (such as it is). *American Scientist*, **68**, 256–67.
Morrison, P., Billingham, J. & Wolfe, J. (eds.) (1977). *The Search for Extraterrestrial Intelligence*, NASA Special Publication **419**.
Raup, D. M. & Valentine, J. W. (1983). Multiple origins of life. *Proceedings of the National Academy of Sciences*, **80**, 2981–4.
Russell, D. A. & Seguin, R. (1982). Reconstruction of the small Cretaceous theropod *Stenonychosaurus inequalis* and a hypothetical dinosaurid. *Syllogeus*, **37**, 1–43.
Simpson, G. G. (1964). *This View of Life*. New York: Harcourt, Brace & World.

Is rape wrong on Andromeda?
An introduction to extraterrestrial evolution, science, and morality

MICHAEL RUSE

The biggest money-maker in the history of the movies starred a little fellow with big appealing eyes, a fondness for candy, and an inability to hold his liquor. I refer, of course, to E.T., the friendly being from outer space, who got left behind when his space ship lifted up prematurely. Audiences all over the world cried at E.T.'s disasters, and cheered at his triumphs. Deservedly, he will surely some day soon get his own star on Hollywood Boulevard.

E.T. is only one element in the general fascination we seem to have today with possible beings from outer space. Successful movie after successful movie plays on this entrancement, as also do some of the longest running television situation comedies.[1] Science fiction books and stories sell inordinately well and there is a constant demand for such old standbys as Superman comic books. On top of all this, government agencies spend our money on searching for possible life either here in our own solar system or further out in the universe. That such searches have so far been unsuccessful seems merely to be a stimulus for renewed efforts and renewed demands for funds.

This fascination with possible life elsewhere in the universe is not a new phenomenon. One finds that the ancient Greeks speculated on the subject, and down through the ages there has been a constant interest in the possibility of life elsewhere. There has also been speculation about its possible nature. Will it be more intelligent than us? Will it worship the same God as us? And above all else, will it be friendly? (See Beck 1972 and Dick 1982 for historical references.)

In the past, philosophers contributed enthusiastically and significantly to the debate about the possibility of extraterrestrial life. Many of the greatest names in philosophy, for instance Aristotle and Kant, had things to say on the subject.[2] Some lesser writers dealt with the topic at great length. Perhaps the most detailed contribution to the extraterrestrial debate – one which denied the possibility of such life – came from the pen of the nineteenth-century British philosopher William Whewell, who argued vigorously that such life is not possible (Whewell 1853). In so doing, Whewell sparked off one of the liveliest controversies of the mid nineteenth century, a controversy which raged until Darwin's even-more-shocking ideas diverted attention elsewhere (Brooke 1977; Ruse 1979*a*).

And yet, despite the history of philosophical interest in extraterrestrial life, and despite the general interest today in such life, modern philosophers stand curiously aloof from the whole debate. There is hardly anything written on the subject. In 1971, as President of the Eastern Division of the American Philosophical Association, Lewis White Beck (1972) used his presidential address to invite philosophers to rekindle their interest in the debate. However, his invitation fell on deaf ears. Somehow, philosophers are altogether too sober a group to discuss extraterrestrials. Dabbling with such entertaining beings as Superman, Yoda or E.T. is too frivolous an enterprise for us. When we turn from serious discussions about the foundations of epistemology and ethics, it must be to even-more-serious discussions of such weighty topics as abortion and reverse discrimination. Science fiction is too childish a pursuit for modern philosophers to take seriously.

It is a pity that the philosophical community did not respond positively to Beck's invitation. *Pace* Dr Johnson, philosophy does not have to be serious and ponderous to be important. Even good philosophy can be fun. Moreover, looking at the extraterrestrial life question can pay philosophical dividends in its own right. At least, this is what I shall try to demonstrate in this essay, for I intend to share with you some thoughts about life elsewhere in the universe.

Why consider extraterrestrials?

I must explain my own interests and the consequent direction my discussion will take. In a funny sort of way – particularly funny for someone who is about to write an essay on extraterrestrial life, I am not myself that interested in whether or not life really does exist out there in the universe. For someone like William Whewell, the possible existence of extraterrestrial life, particularly extraterrestrial intelligent moral life, was rightly a matter of some personal concern. As a practising Christian, he felt that, were worlds other than ours populated with intelligent beings, then necessarily they would have to have some kind of relationship with God. And this definitely dilutes any special unique relationship that we humans can claim to have with God.

Indeed, the possibility of such beings elsewhere in the universe raised for Whewell all sorts of horrendous questions about the possible need for Jesus to come down from heaven and be crucified for their sakes, as well as for ours. One has this dreadful thought that on Friday, every Friday, somewhere in the universe Jesus is being hanged high for someone's sins.[3]

Whewell's critics, who argued that there is indeed life elsewhere, had no fewer personal concerns than he. The most strenuous objector was the Scottish educator, scientist, and ardent natural theologian, Sir David Brewster. Author of a vitriolic, passionate work with the glorious title, *More Worlds than One: The Creed of the Philosopher and the Hope of the Christian* (1854), Brewster was desperate to show that no part of God's creation is without purpose. Thus Brewster felt that he had to prove that everywhere, throughout the universe, there is teeming life. This extends even to our sun, despite the fact that, heat problems aside, any human on the sun's surface would weigh somewhere in the region of two tons, because of gravity.

My worries are not the worries of Whewell and Brewster. My religious beliefs need no proof or disproof of the existence of beings elsewhere in the universe. Indeed, from a personal point of view, I am quite indifferent as to whether or not there is life elsewhere in the universe. I find this life of ours quite enough to handle! Furthermore, although I am sure that a philosopher could rightly enter the dispute about the probabilities of such life existing elsewhere in the universe, I shall not really do so in this article. Some well-qualified scientists today argue that not only is extra-terrestrial life possible, it is indeed probable (Billingham 1981). Other equally well-qualified scientists argue that extraterrestrial life is most improbable (Hart & Zuckerman 1982). I suppose that if I were pushed, I would admit to a sneaking suspicion that, possibly, extraterrestrial life does exist. But, I must agree that often calculations of the possibility or probability of extraterrestrial life put one in mind of the engineer's way of solving problems: namely, think of a number, double it, and the answer you want is half the total.

All too often, in the face of ignorance, we make assumptions which sound reasonable, but which are totally without justification. Suppose you are told that there are some hundred million possible sites of extraterrestrial life in the universe. Suppose then the suggestion is thrown in that, on one in a thousand of these sites, such life has in fact evolved. That sounds reasonable enough. Conservative even! What is one in a thousand? But without evidence (usually lacking) it is quite without foundation. Why not one in ten, or one in ten thousand, or one in a million, or one in a hundred million? Unless one has either a theory to back up one's hypothesizing, or some empirical evidence to show that one's probabilities have some possibility of being reasonable, any supposed probabilities that one pulls out of thin air are no more than that.[4] Thus, I conclude that perhaps life exists elsewhere in the universe, perhaps it does not. Perhaps such life is intelligent, perhaps it is not. I just simply do not know.

Why then, you might ask, should one professing so little interest, and even less knowledge, presume to write on extraterrestrial beings? The answer is that, by so writing, we can throw interesting light on ourselves. Moreover, egocentrically, I must confess to an extreme interest in human beings. Hence, my aim in this essay is to consider extraterrestrial beings – most particularly to consider their possible natures – hoping thereby to understand more fully the factors which make human beings what they are now. I am particularly interested in human claims to knowledge, both in the realm of epistemology (what Kant would call pure reason) and in the realm of ethics (what Kant would call practical reason).

Thus, supposing that there are beings elsewhere in the universe, the questions I ask are the following: How did they come to be? Was it through some form of evolution, and if so, what form? What are they like? How then do they compare in origins, similarities and differences with us human beings? What does this tell us about ourselves? Specifically, what do we learn about our intellectual and moral capacities, potentialities, and limitations?

Evolution?

Let us assume that somewhere in the universe we have a planet roughly like ours, together with the approximate combination of chemicals and conditions that obtained on our planet some $3\frac{1}{2}$ or so billion years ago. In other words, let us assume that we have a planet like earth, after it had cooled down somewhat, but before life had yet appeared. I have admitted that I do not know if this is at all a reasonable assumption. Since I am carrying out a thought experiment, it is enough that the assumption be logically possible.[5]

As a determinist and a mechanist, I certainly believe that, given the identical combination of chemicals and conditions as prevailed here on earth, life would necessarily appear in exactly the same way. But even thought experiments cannot ask this much – at least they shouldn't. I'm just assuming that we have more or less the same chemicals and conditions. And this being so, we certainly cannot say definitively that life would appear on our hypothetical planet. We have neither established theory nor strong empirical evidence to let us say this much.

However, today, unlike just a few years ago, we are far from completely ignorant about the coming into being of life.[6] We have some very suggestive empirical evidence showing how the components of life could be formed naturally from non-life substances. I refer in particular to the work, in the 1950s, of people like Miller, who showed that amino-acids can be formed from inorganic substances without direct human intervention. Furthermore, recently a number of very stimulating theories have been developed about how life could develop naturally from these components (Dickerson 1978; Schopf 1978). For instance the Nobel prize-winner Manfred Eigen has devised a theory showing how ribonucleic acid (RNA) could be formed without human or other intelligent direction (Eigen *et al.* 1981).

Hence, it is no longer true to say that we are totally without any knowledge whatsoever about life appearing naturally. Nor is it reasonable today to conclude that, given a planet much like ours, life could never have appeared. But, again, I will not become too bogged down with facts! If you think that, given a planet like ours, life would not appear, then give me a planet like ours with life. Either way, let us move on. (With my cavalier disregard for facts, I am even less concerned with whether life could have occurred under drastically different conditions and combinations of chemicals: for instance, whether one could have some sort of ammonia-based life. If indeed such alternatives are real possibilities – I was tempted to say 'live possibilities' – the points I want to make as this essay progresses will still hold valid. And if they are not real possibilities, we have saved time by not being side-tracked by them.)

Let us suppose then that life does develop in some primitive sort of form. I am deliberately avoiding heady philosophical discussions about what we mean by 'Life.' I am simply assuming, backed by suggestive experimental and theoretical work mentioned just above, that we have primitive organisms which are relatively stable. Moreover, and this is crucial, they *reproduce*. They do not just sit around in splendid isolation, contemplating their DNA. They multiply.

At this point, big, obvious questions intrude. Will we get evolution? And if so, what will fuel this evolution? Here on earth, we do have evolution, and *the* major mechanism of change is 'natural selection.' More organisms are born than can possibly survive and reproduce. In the consequent 'struggle,' some prove more successful or 'fitter,' by virtue of their special characteristics. This leads to a kind of natural winnowing or 'selection.' Given enough time, we have evolution. But would the terrestrial mechanism of natural selection be as important on our hypothetical planet, as it undoubtedly has been down here on this real planet?

There will be evolution, and natural selection will be crucially important. Why can one say this? The answer lies less in evolution itself, and more in the natural selection which will in turn lead to evolution. Natural selection is not a tautology, as some critics claim.[7] However, it does rest on basic properties of organic matter, and, under the properties already proposed for our primitive life, it is hard to suppose that it would not apply up there to our hypothetical planet, as it has down here to our real one.

Consider: We are assuming that we have organisms which do reproduce, and I take it we are assuming that they do not simply reproduce one of their kind and then die out. Rather we have organisms which multiply. Obviously, what we are going to have is an explosion in numbers, as we have a Malthusian geometric increase. Eventually, fresh living space and additional nutrients required for survival will be exhausted. What then? On our hypothetical planet, there will be a struggle, as there is down here on our real planet. If all of the organisms are absolutely identical, then success will be a matter of chance. There will be no selection. But, if there are differences, and everything we have learnt about life here on earth suggests there will be differences, then the slightest edge will lead to success in the struggle. We shall have natural selection on our hypothetical planet.

Moreover, this natural selection will lead to evolution. That is to say, one will get an ongoing change in the organisms on the planet. This is not to say that all of the organisms will be changing all of the time. There could be long periods with little or no change, as occurs on earth (Stebbins & Ayala 1981). But, given enough time, one expects change to occur. This is an expectation backed by theory and by terrestrial empirical evidence (Ruse 1982a).

But what does it all lead to?

What route would evolution take on a hypothetical planet, and what effects would it lead to? In many respects, it is simpler to say what sorts of things we should not necessarily expect to find. Most particularly, there is absolutely no reason to think that we are going to have an inevitable progression from primitive forms to intelligent beings. As mentioned earlier, here on earth life first appeared about $3\frac{1}{2}$ billion years ago. But, until just over a half billion years ago, such life was very primitive. It is only in the last 600 million years or so, that we have had the development of sophisticated organisms and plants (Valentine 1978).

Hence, if anything, the empirical probability seems to be that life on our hypothetical planet will remain at a very crude level. Success in the struggle for existence is not necessarily a question of being brilliant. Rather, it is a question of being a better reproducer than your neighbor. If your neighbor is burdened with brains, whilst you are little more than a fancy sperm sac, then it could well be that you will reproduce far more efficiently than he, she or it.

Moreover, although a number of hypotheses have been put forward, no one has yet given absolutely convincing reasons why, at the beginning of the Cambrian or thereabouts (i.e. 600 million years ago), there was an explosion in life forms, including sophisticated life forms (Raup & Stanley 1978; Futuyma 1979). Hence, we have neither theory nor empirical evidence to suggest that life will necessarily evolve into something more than primitive single-celled organisms – however long a planet may exist. There is certainly no guarantee of birds and fish and reptiles and mammals and plants and everything else that makes life recognizable for us. There is absolutely no guarantee of races of Yodas or E.T.s or Supermen. If anything, we should expect the persistence of low-grade life.

What then can we say about the effects of natural selection on our hypothetical planet? The most important thing about natural selection is that it does not just lead to any kind of change. It brings about adaptation. That is to say, natural selection makes for characteristics which aid their possessors in life's struggles. Moreover, these are characteristics which help each organism individually. Such features are known as 'adaptations.' Hands, eyes, ears and noses are all organs which help one to survive and reproduce. These are all adaptations (Williams 1966; Lewontin 1978; Mayr 1982).

Hence, on our hypothetical planet, we expect that natural selection will bring about adaptations. Moreover, if most people have got one particular characteristic, helping survival and reproduction in one particular way, it often pays to have different characteristics, helping survival and reproduction in other ways. Natural selection usually leads to diversity. If everyone is trying to eat one foodstuff, then, if there is another available foodstuff, a liking for this second foodstuff is clearly to one's advantage. I expect, therefore, that on our hypothetical planet, we would find different forms of organism (Dobzhansky 1970).[8]

Whether we should find sexuality on our planet is an interesting question, but I am not sure that we can really answer it that definitively.

There are almost as many theories on the evolution and maintenance of sexuality, as there are writers on the topic (Ghiselin 1974; Williams 1975; Maynard Smith 1978; Trivers 1983). There does seem, however, to be strong evidence to suggest that sex has evolved more than once here on earth, so perhaps it might evolve elsewhere also. If indeed we do have sexuality, then probably we should have species: groups of reproductively isolated interbreeding organisms. There is still much controversy about how speciation actually occurs, but it does seem to come about as a consequence of sexuality. As soon as organisms start breeding with one another, then things happen which stop that breeding from being ubiquitous (Dobzhansky 1951; Mayr 1957, 1963; Grant 1971, 1981; Sokal 1973; Van Valen 1976; Endler 1977; White 1978; Levin 1979).

Let us move on now towards questions about human evolution. I have said, and I am sure truly, that there is absolutely no guarantee of an upward progression on our hypothetical planet to intelligent life forms. On the other hand, evolution is constantly trying to exploit new unused domains. As one area, for instance the sea, becomes used up and occupied, then we find evolution pushing organisms into unused areas, such as land, and as that is used up, then we find evolution towards other areas, such as the air (Valentine 1978; Feduccia 1980). And similarly, in our own case, we have seen an evolution from a more biologically oriented sphere to one which involves more of a cultural dimension, and which is less connected absolutely with brute nature. There was nothing inevitable about this development, but it did happen nevertheless. Furthermore, notwithstanding all of the reservations, intelligence is not something terribly counter-intuitive, given natural selection. One simply has the constant need to better oneself, and bettering oneself involved becoming a little more sophisticated. Intelligence is the epitome of sophistication (Johanson & Edey 1981; Lovejoy 1981).

What I suggest, therefore, is that through a process of natural selection, we might possibly have the evolution of more sophisticated organisms on our hypothetical planet, and these could even reach the realm of intelligence. Certainly it has happened once, namely here on earth, so I do not think it is empirically impossible that it happened elsewhere. But I do emphasize that such evolution of intelligence is not a necessary consequence of life appearing: not at all. (One important question, which I will leave undiscussed for the moment, concerns possible connections between sex and intelligence. I will look at it later.)

To keep our discussion going, let us suppose that natural selection does lead to some sort of intelligent race or species of organism on our planet. (At least once here on earth, we had co-incidentally two species of semi-intelligent organism: *Australopithecus africanus*, and *Homo habilis*. Members of neither species were as intelligent as members of *Homo sapiens* are.) The question which arises now is just what sorts of knowledge systems would these extraterrestrial intelligent beings have? In the next section, I turn to the question of epistemology. That is to say, what sort of knowledge would they have of the world around them? For instance, would they know of Newtonian gravitational theory, and of like discoveries of the human mind?

Extraterrestrial empirical knowledge[9]

There is an obvious, even if understandable, fallacy which occurs in popular science-fiction movies. This is the making of the extraterrestrial beings as recognizably human, particularly with respect to their senses. Admittedly, Superman had supposed powers which are beyond us; but, generally, the extraterrestrials are given those very senses which we have, often with the same intensities. Thus, for instance, E.T. had two eyes, just like humans, as well as a human sense of taste (otherwise, why was he so fond of Reese's Pieces?), together with a human sense of touch and smell and so forth.

But we can certainly say from our experience here on earth that there is no reason at all why an intelligent denizen of our hypothetical planet should have precisely those senses that we have, in precisely the way that we have. For instance, certain snakes can strike their prey in the dark, by virtue of having incredibly sensitive heat sensors on their heads. I see no reason at all why an extraterrestrial being should be blessed with a human sense of sight, rather than with sophisticated heat sensors. Again, there are more ways of seeing than simply the human ways. Spiders and other animals have evolved sense of sight quite independently from the mammals, so it could be that our imaginary beings would see parts of the spectrum different from those which we see.

Also, it is well known that many animals – fish, insects and others – communicate by chemical means, through so-called 'pheromones.' To give you an idea of how efficient this method of communication might be, let me quote from a discussion by E. O. Wilson, an expert in the nature and use of pheromones in the organic world.

'The amount of potential information that might be encoded [by the use of pheromones] is surprisingly high. . . . Under extremely favourable conditions, a perfectly designed system could transmit on the order of 10,000 bits of information, an astonishingly high figure considering that only one substance is involved.

Under more realistic circumstances, say for example in a steady 400-centimetres-per-second wind over a distance of 10 metres, the maximum potential rate of information transfer is still quite high – over 100 bits a second, or enough to transfer the equivalent of 20 words of English text per second at 5.5 bits per word. For every pheromone released independently, the same amount of capacity could be added to the channel capacity.'
(Wilson 1975, p. 233.)

Clearly, we humans hardly begin to exploit the way in which one can handle information. To think that extraterrestrials will use all, and only our, five senses is unjustifiably anthropomorphic in the extreme. Intelligent denizens of our hypothetical planet could be blind snakes, rather than human-sensing fellows like E.T. Do not expect souped-up humans like Superman.

I suppose, however, that if extraterrestrials are to do anything with their intelligence, they will have to have some ability to manipulate their environment. Perhaps indeed, intelligence and an ability to manipulate the environment go hand in hand. (If you will forgive the metaphor.) Certainly, the history of human evolution seems to imply that physical and mental evolution are connected, although now it is believed that we pushed towards our present physical state somewhat before we pushed towards our present mental state (Johanson & White 1979; Johanson & Edey 1981). But, I do not see in principle that natural selection lays down absolute rules for the way in which the evolution of intelligence must take place. It is simply that in order for the evolution of intelligence to occur, one must in some way be able to put it into practice. One therefore looks to feedback between the ability and the means of utilization. One is not going to develop straight into a philosopher, able only to think and not to act.

Will our extraterrestrials be conscious?[10] Frankly, I am not quite sure how you would answer this question in any definitive way. We know, each and every one of us, that we are conscious. We are as sure of consciousness of our fellow human beings as makes no real difference. We are equally sure that stones are not conscious and fairly certain that lower organisms like oysters have little by way of consciousness. There are those who have denied that any animals, other than humans, have consciousness. René Descartes was one. I suspect, however, that most people would agree that consciousness is at least in some sense linked with brain power, and that it is shown through physical actions, behaviors, responses, and so forth. This being so, most people would allow that the very highest organisms, (other than humans) have some rudimentary form of consciousness.

Certainly, it seems a little odd (selfish?) to deny that chimpanzees and gorillas have consciousness of any kind. They engage in behavior which goes along with what we associate with consciousness in our fellow humans, and indeed in ourselves: gorillas and chimps are able to manipulate tools, make discoveries, behave in social ways, and so forth. They do not have verbal speech; but, it is arguable that actually being able to talk is a necessary condition for any form of consciousness whatsoever, especially since there is increasing evidence that the higher organisms can communicate in non-verbal ways (de Waal 1982).

Let us grant that animals other than ourselves are conscious. At least, let us grant the weaker premise that an organism as sophisticated as we in brain power could be conscious. Hence, there is no good reason to deny the possibility of consciousness in extraterrestrial beings. However, oddly, I am not sure whether the possible consciousness of our extraterrestrials is all that important, or all that significant. At least not for us! I see no reason why extraterrestrials should not behave and react in just such a way that we would associate with reasoning, whether or not the kind of 'ghost in the machine' so beloved of dualists actually exists. Put matters this way. In the *Star Wars* series, C3P0 and R2D2 are machines. Luke Skywalker and Princess Leia are human-like. Chewbacca and Yoda are intelligent non-humans. Does one want to say that any or none of these is conscious? Does it matter?

We come now to a second, and rather more interesting, question. What sort of thinking beings would our extraterrestrials be? What sort of consciousness would they have, if they had consciousness? Clearly, in many important respects, our extraterrestrials are going to think in ways very different from those of us humans. At least, we cannot expect that they will not. For instance, if their senses are essentially chemical-using, then they are going to have a whole range of sensations totally unexperienced by us. Conversely, the sensations experienced by us are going to be unexperienced by them (Grossman 1974).

At the very least, what our extraterrestrials take for granted will be theoretical concepts for us, and vice versa. As I pointed out, their metaphors, similies and all of those sorts of concepts are going to be quite different. One should not underestimate this. There seems to be something of a tendency to think that analogies and metaphors are useful aids, but theoretically dispensible; that they are not things which enter into real thought, in particular into science. However, in common with a number of other philosophers of science, I would dispute this. Metaphor plays a very important role in science, and to eliminate it would be to alter science very drastically (Black 1962; Hesse 1966; Ruse 1973*a,b*, 1977, 1980*a,b*, 1981*a,b*).

Indeed, I am not at all sure that one could eliminate metaphor entirely from science. Therefore, to say that our extraterrestrials could grasp what we would think of as Newtonian mechanics is dubious. For instance, for us an important part of Newtonian mechanics is a notion of 'force.' This is clearly something we get through the idea of actually physically stretching and pressing, and so forth. However, if one did not communicate through touch at all, or at least only in a very minor or unimportant way, but if one communicated through some sort of chemical sensing, then I doubt that the notion of force could play the same role in extraterrestrial mechanics that it plays for us. Conversely, some other metaphorical notion would be at the centre of their theory. That is, assuming they have an equivalent theory.

Again, in company with other philosophers of science, I take the notion of a 'problem' to be very important (Laudan 1977; Ruse 1980*b*). Scientists do not just go out and discover the world – any old part of the world. They get stimulated by problems. Newtonian mechanics is obviously a function of our awareness of the heavens and of gravity (to name but two factors). What if we were blind fish? Would we feel the need for Newtonianism? Certainly not in the same way. Astronomy would be less important. Hydrodynamics would be more important.

But is there not some universal knowledge?

However, you might think that, despite all differences between extraterrestrial science and human science, there would nevertheless have to be a basic underlying similarity between any kind of human thinking and any kind of extraterrestrial thinking. After all, $2 + 2 = 4$ on (a planet on) Andromeda, no less than it does here on earth. Squares are squares and not circles on Andromeda, just as they are here on earth. And, whether or not we can or do think of 'force = mass × acceleration' on Andromeda in human terms, it holds up there equally as much as it does here. And the same is true of Boyle's law, Snell's law, and all the rest. Furthermore, if, as I have just been suggesting, extraterrestrials evolved through natural selection, then natural selection is no less a reality on Andromeda than it is down here on earth.

Hence, whether or not the periphery of science is changed, there is something underlying which is the same. There is, therefore, a universality to what can and will be known, despite the differences. The situation is analogous to that seen by someone like Chomsky (1957) in linguistics, when he argues that there is a universality to language, a so-called 'deep structure,' despite the many surface differences there are between (say) English and Japanese.

Let me make two comments about this point. First, even given the amount that the critic would allow, I think it would be terribly easy to underestimate the grave differences which could still exist between us and extraterrestrials. We could have to battle all the way for even the most primitive communication. We know full well how difficult it can be to understand the minds of folks in cultures different from our own. How many Westerners, for instance, would truly want to say that they can understand oriental inscrutability? But, if there are so many difficulties encountered in communicating with, and fully understanding, our fellow humans, think how much more difficult it could be to try to communicate with extraterrestrials.[11]

Certainly any kind of profound understanding of extraterrestrial culture could be forever out of our grasp. Conversely, any understanding of us could be out of the extraterrestrials' grasp. And, do not think that breakdown in communication would occur just in the arts, although how one would explain Wordsworth's poem about daffodils to a blind watersnake I simply do not know. As I have argued above, we could have a great deal of difficulty communicating many, many scientific ideas, given the extent to which they are permeated with our human values and nature.

The second comment is that, despite all difficulties of communication, I think that there would remain some common ground in human and extraterrestrial knowledge. Essentially, what I am arguing is that there is a common world out there, one to which we and our extraterrestrials will have adapted. Without such an assumption, everything I have said so far falls away. Hence, I am committed to a realist position. (The sense of 'realism' will be spelt out shortly.)

The world is not just a figment of our imaginations. Rather, we are organisms who have adapted, through natural selection, to deal with external realities. Moreover, I have no reason to doubt that really $2 + 2$ does equal 4. Why? Because I have no reason to, that's why! It is simpler to assume that a being which recognizes that $2 + 2$ does equal 4 is better adapted than one which does not, if indeed $2 + 2 = 4$ and not 5.

Again, I have no reason to believe that circles are not really circles and are really squares. Why? Because, it seems more sensible to assume that an animal which has developed adaptations enabling it to deal with squares which really are squares rather than circles is in some sense better off than an animal which has not. Furthermore, I see no reason to think that on Andromeda, or wherever, $2 + 2$ will not equal 4, or that squares will really be circles, or that the fire will not cause burning, or that more generally the same cause will not lead to the same effect. And I see no reason to doubt that Andromedians will be better adapted if they (like us) can distinguish circles from squares, and can learn that fire 'causes' burning. [But, note what I have to say later about causation.]

In short, I am assuming that there is going to be an underlying similarity between the world humans live in and any world elsewhere. Additionally, extraterrestrials, just like humans, will have learned to cope with their world and to recognize the underlying patterns which are common to the whole universe. I would even go so far as to say that if there are good adaptive reasons for believing in the existence of a God, namely it promotes a certain group stability and cohesion and so forth, then perhaps extraterrestrials will believe in God (Wilson 1978). But, remember the incredible varieties of beliefs in God down here on earth, with the Hindu notion of God bearing but the faintest family resemblance to the Christian notion. Even if we met extraterrestrials, I am not at all sure that we would be able to understand what their notion of God was, or even that they had one.

My key position, therefore, is that there will be some underlying basic premises held in common by humans and extraterrestrials. I have in mind such things as the basic premises of logic, of mathematics, of causality and so forth. In other words, I am thinking of the kinds of concepts that Kant identified as analytic, and perhaps also the synthetic *a priori*. Thus far, extraterrestrial epistemology will be similar to human epistemology. [Again, note the qualification I make later, at the end of the next section, about the 'reality' of such things as causes.]

Ultimate reality?

There is a major question which must be raised and looked at fairly. I am suggesting that organisms are a product of natural selection, and that they have senses because these are adaptive. What this means is that having senses enables organisms to survive and reproduce better than if they did not have them. Clearly, this is no guarantee of ultimate truth in any sense. It is rather a pragmatic position (Quine 1969). One has certain senses because these work. But, since humans are themselves organisms, cannot the argument be turned right back on everything that I have just been saying? What guarantee have I that any claims I make about natural selection itself are really true? What guarantee that my claims really tell us of that which is out there, beyond human fallibility and adaptive need? Is natural selection absolutely true of ultimate reality, even though we ourselves learn of it through organs themselves fashioned by selection? Indeed, can we properly talk about ultimate reality, as opposed to common-sense reality we experience all the time – an everyday reality to cope with which selection gives us adaptations? To use the terminology of the philosopher Hilary Putnam, can we talk of 'metaphysical reality' as opposed to 'internal reality' (Putnam 1981)?[12]

Put matters this way. My human senses, my human powers, are themselves products of natural selection. Therefore, everything that I am saying could be totally mistaken (Trivers 1976). My claims could be simply things I believe, because I am a more efficient reproducer if I believe them, rather than otherwise. In short, apparently, ultimately I have no right to say even that 2 + 2 really equals 4 or that there really are chairs and tables where I think there are, or that the fire always causes pain, or anything else. I think these things hold true because these beliefs enable me to survive and reproduce. But what ultimate reality is like, I cannot say. Nor – and this is the punchline – can I say that ultimate reality for extraterrestrials will be the same as it is for us.

This is a worrisome objection, but we can go some way towards answering it. Moreover, the distance that we can go provides all the answer that we really need. First, note that we are not entirely helpless or uninformed about the world out there. Certainly, we are not uninformed about the effects of different senses, or of senses alien to us. At the personal level, we ourselves have several senses, and the interesting thing about these senses is that, although they give different information, they do nevertheless cohere on the reality of the world.

I think, in particular, of the sense of sight and the sense of touch. It is well known that the proper objects of sight and the proper objects of touch (as Berkeley used to call them) are not the same. A blind man who knows a square from a circle by touch, cannot immediately distinguish a square from a circle when once he is given his sight.[13] Nevertheless, the blind man can recognize the square from the circle just as the man with sight can. He knows that he has two different objects, and that the one is what he would call a 'square' and the other a 'circle.' Moreover, these are just as stable and usable as are the 'squares' and 'circles' of the man of sight. Hence, although different senses give us different information, they do cohere on such things as differences between squares and circles, as well as on 2 + 2 = 4 and so forth. Therefore, the fact that we have different senses does not mean that we cannot or do not agree on the same basic underlying principles.

Moving now from ourselves to other animals in this world of ours, note that chimpanzees seem to avoid objects in their path, just as much as humans do. (Rather better, in fact!) Similarly, inasmuch as chimpanzees can manipulate and so forth, they do it in essentially the same sort of ways that humans do and for the same sorts of reasons. Again, moving now beyond straight sensation to the ways in which we process it, we find that chimpanzees learn to associate cause and effect, in the same ways that humans do. For instance, a fearful sound sets up an expectation in a chimpanzee just as it sets up an expectation in the human (de Waal 1982).

What I argue, therefore, is that we do not get a total breakdown between different organisms, or between organisms of different species, with respect to the fruits of sensation. Nor do we get total breakdown with respect to the basic principles that I suggest might underlie all epistemology, including human epistemology. This is not to say that chimpanzees or lower organisms are as aware of these principles as humans. In fact, I suspect that we are the only organisms here on this earth consciously aware of the kinds of underlying principles I am talking about. But what I am saying is that they do seem to be shared by organisms of different species.

Even when we move to organisms which use entirely different senses, like those organisms relying on pheromones, we still find that there seems to be the same basic consistency between them and us, and other organisms using senses like ours. For instance, snakes using heat sensors strike at objects which we can see. They do not strike at invisible prey, and then slither through that which we sense as solid. And the same points hold for animals using pheromones. We see the objects or we feel the objects; they sense them chemically or some such way (Wilson 1975).

In other words, there seems to be a basic underlying consistency with respect to reality, and to the primary properties of reality, which transcends different senses. There seems, therefore, no reason to deny that the same should be true of our extraterrestrial beings. Although their ways of sensing, and so forth, could be quite different from ours, the evidence from ourselves and from our fellow earth-organisms is that reality is the same for all, and that our various senses are all attuned to react to its fundamental properties. Hence, at this level, one can say confidently that extraterrestrial beings will share with us certain fundamental, underlying principles. Whether they will have the precise awareness of them that we have is another matter. But the principles will be shared nevertheless.

Having said this much, however, there is still the underlying worry that, in some ultimate way, we may just be entirely mistaken. What is shared by humans and extraterrestrials is still the reality revealed by the organs of selection (Putnamian internal reality). If, as I argue, our senses and powers are adaptations to deal with the world, the critic might yet object that there is no absolute guarantee that we are not systematically deceived about some aspects of ultimate reality (Putnamian metaphysical reality).

Descartes supposed an evil demon who misled us. Perhaps natural selection misleads us 'for our own good' (Trivers 1976).

At this point, I simply have to throw up my hands and admit that one can set up the situation in such a way that, ultimately, I have no answer. But, note that this supposition says nothing to refute what I have just argued about the similarity between ourselves and extraterrestrials. I see no reason to believe that Descartes' evil demon would not deceive them as well as us, or that they would not be subject to the same kind of natural-selection-caused distortions about ultimate reality that we are.

However, I do not want to end on a negative note. The reader should not think that, because I admit we are limited, we can say nothing about ultimate reality. And that such ultimate reality could be of a nature that, if perhaps we only had extra senses, we could find out more, and that we could perhaps find out that $2 + 2 = 5$, and why it is biologically advantageous for humans to believe $2 + 2 = 4$. And that perhaps extraterrestrials do have the needed extra senses, and thus can find out more than we do, even if they too in the end are necessarily limited by their abilities.

To counter this worry, let me say the following. First, there is something awfully fishy about this line of argument, if only because we are pre-supposing the truth of natural selection to prove that it might not be true. If in fact the real world is in no wise as we picture it, then it hardly makes sense to say that natural selection is systematically deceiving us for our own good. All we can do is remain silent before the imponderable – or, more precisely, before the inconceivable. To say that we are totally deceived by reality, just does not make sense. This is not to say any 'deception' is ruled out. I suppose, for example, we might think a certain colour identical with another – although I am not quite sure why this should be a product of selection and not simple insensitivity. Perhaps, more plausibly, we might say that we could all be deceived about the existence of God – although, please do not take this comment as a plea for atheism. Apart from anything else, if it were argued that the use of the term 'deception' stretches metaphor too far in this case, I would be quite sympathetic.

Secondly, backing up the first point, because our powers are limited, or rather, because our powers are what we have and just a product of natural selection, it does not follow that it is reasonable to suppose that reality is other than what we think it is. To say that our powers are limited and that we can say no more is just to say that it is inconceivable that ultimate reality is other than what we think it is. And to say that things are inconceivable is not in any sense to say that they are reasonable or probable or another option. They are literally inconceivable, and one can say no more than that. Hence, I find myself untroubled by the thought that ultimate reality might be other than it is, or that extraterrestrials might have deep insights into this reality – insights hidden from us. They might perhaps sense colors we cannot sense. I doubt they think $2 + 2 = 5$, either because it is true or because natural selection finds it worth while systematically to deceive them. And, even if it did, I am not sure how we would ever find out.

Concluding this part of the discussion, therefore, I argue that there could well be radical differences in the epistemology of humans and extraterrestrials. But inasmuch as we can both think about the world at all, there will be certain similar underlying thought patterns. And we share our powers of comprehension, because we are both products of the same process of evolution. Were a Humean to object to my analysis on the ground that, although objects may exist 'out there', such things as causal connections do not refer to real powers, but are mind-supplied to deal with reality, I would feel sympathy. However, even if we only have constant conjunction in reality (and I am inclined to think that is all we do have), my position is simply that thinking in terms of cause and so forth is an adaptation for dealing with reality – imagine if we did not associate fire with burning – and it is inconceivable that extraterrestrials would not share such adaptations.[14]

You may think that, in this sense, selection is 'deceiving' us. But apart from the fact that David Hume was not deceived, I have already hinted that emotive words like 'deception' are probably not really appropriate here. We are not consciously being led into falsity. (I would like to think I am arguing for a Kantian solution to Hume's problem, but one which is infused by Darwinian insights.)[15]

Extraterrestrial morality

I come now to the most interesting questions of all. What about right and wrong, likes and dislikes? Will inhabitants of our hypothetical planet have any kind of moral code? If so, will the code be anything in any way recognizably like our own? And, most particularly, how would such extraterrestrials feel and behave towards us humans if we were ever to meet them? Perhaps, I should also ask what kinds of moral obligations we humans would have in return towards extraterrestrials?

If you are a Christian or a Platonist or some such thing, then you believe in an absolute moral law. This being so, then whether extraterrestrials exist or not makes no difference to ultimate moral reality. There is one ultimate moral rule for us all, whether we live here on this planet or somewhere else, and that is that. However, if you are like me, you are nowhere like as convinced of such ultimate rules. Matters therefore become rather more complex and more problematical. And, of course, even if you believe in an absolute moral code, you still have questions to do with the reactions that beings very much unlike ourselves would have towards this code, and the obligations which would extend to them.

What follows if you take the kind of approach that I have been taking in this essay? What happens if you think that evolution through selection will be the causal key to intelligent life elsewhere in the universe, as well as down here on earth? Until recently, I think most people would have felt that there was little, if anything, that one could say about morality. There was a strong feeling, even by evolutionists themselves, that evolutionary thought, particularly evolutionary thought centered upon natural selection, had little or nothing to say about the evolution of morality. It was felt that evolution promotes little more than an unbridled selfishness where all organisms fight for their own interests entirely to the detriment of any other organisms. In other words, one has an unrestricted 'nature red in tooth and claw.' One has social Darwinism, where the weakest go to the wall and the strongest triumph (Dobzhansky 1962; Flew 1967; Huxley 1901).

This is obviously all directly opposed to any kind of morally governed society, where one has kindness and friendship and altruism. Most people, therefore, including evolutionists, felt that somehow morality was a function of non-biological culture or some such thing. Morality and evolution are opposed, not complementary or in any way in harmony.

Recently, however, there has been a revolution in evolutionary thinking about the beginnings of morality. More accurately, we should say that there has been a return to the thinking of Charles Darwin himself, for in the *Descent of Man* Darwin was quite adamant that conventional morality can be given a selectionist backing (Alexander 1971, 1979; Trivers 1971; Wilson 1978; Ruse 1979*b*; Murphy 1982; although see also Singer 1981).

There are a number of causal models which are suggested in explanation of the way or ways in which our sense of morality could have evolved. One of the most developed is that which applies to relationships within the family. It is argued that there are good evolutionary reasons why people should behave in an altruistic, moral way towards relatives. In particular, we all share the same basic units of heredity, the genes. Thus, inasmuch as one helps relatives to reproduce, one is thereby helping oneself to reproduce, by proxy as it were. This mechanism, which has been labeled 'kin selection,' is thought to be responsible for a great deal of the altruistic behavior which we show towards the people around us, for instance, our children and so forth. Obviously, help to relatives does not have to be confined to the actual mechanics of reproduction. Financial and other indirect support is covered by kin selection (Hamilton 1964*a*,*b*; Trivers 1972; Dawkins 1976; Maynard Smith 1978, 1982; Oster & Wilson 1978; Wilson 1978).

What about dealings with non-relatives? Here also evolutionists (of the Darwinian ilk) argue that natural selection was important. In particular, it is argued that much morality evolves as a product of a kind of enlightened self-interest, or, as evolutionists call it, 'reciprocal altruism' (Trivers 1971; Alexander 1979). It is suggested that, inasmuch as one behaves in a friendly altruistic way towards other human beings, so also will they be prepared to behave in such a way to you. Moreover, those humans who are helped by others are in a better position to survive and reproduce than those who are not. This, even despite the fact that sometimes the recipients of help put themselves out for other people.

You might object, with Immanuel Kant, that true morality occurs only when one is a totally disinterested participant. That is to say, morality begins only when one does something without any hope of favor or reward. However, evolutionists anticipate this objection (Alexander 1974, 1975; Ruse 1979*b*). They argue that the evolved sense of morality in humans does not necessarily involve conscious manipulation or calculation of possible return. In fact, evolutionists argue that one will probably function most efficiently when one has no hope of return at all. In short, biological functioning is maximized when one is acting solely because one thinks that one ought to act in such a way. In other words, one works most efficiently when one is acting morally.

On the basis of the above and related models, evolutionists today argue that the human sense of morality is something which is a product of evolution, no less than the hand or the eye. Our sense of morality is an adaptation. We survive and reproduce more efficiently with it, than we do without it. In the past, those people who lacked a sense of morality, simply tended to be ostracized or otherwise at a disadvantage. Thus, they failed to survive and reproduce as efficiently as those with a sense of morality. This is not in any way to say that that which has evolved is morally good. In other words, this is not some form of crude social Darwinism, either endorsing the process of natural selection as a moral good, or the products of natural selection as moral goods. It is not to say that nature red in tooth and claw is an unqualified good. It is rather to say that the sense of morality which we have is a product of evolution.

Now, how does this all bear on the question of extraterrestrials? Remember, we have concluded that extraterrestrials will, like ourselves, be a product of a process of natural selection. My surmise is that the correct answers will come at two levels, just as our answers to extra-terrestrial epistemology came at two levels. Moreover, I think that the two levels answering to morality will bear significant similarities to the two levels answering to epistemology. There, it will be remembered, it was suggested that sensations, models, metaphors, analogies and so forth, will probably be quite different for extraterrestrials as opposed to humans. However, at another perhaps deeper level, there will be signifi-cant similarities, particularly with respect to mathematics, logic, prin-ciples of causality, and so forth. I would suggest that in the realm of morality, at a more surface level, we will find very significant differences. But, perhaps, at a deeper more basic level, there will be similarities between extraterrestrial morality and human morality. Let me explain.

Subjective elements in morality

At the top surface level, in human morality you find all sorts of subjective and variable elements. Thus, for instance, we find many differences between different societies. Take eighteenth-century India: people in that time and place practiced and strongly endorsed 'suttee'. This was the custom whereby widows would voluntarily perish in the funeral pyre of their husbands. Obviously, such a practice is totally alien to Western customs and morality. In fact, we think that widow sacrifice is totally immoral.

Again, we find differences between practices in Africa today and what we think moral in North America. Recently, there has been some controversy about the practice of female circumcision. This involves the removal of a significant part of the external female genitalia. Inhabitants of many parts of Africa think that this is a good thing to do, and that one ought to do it if one is to be a morally respectable member of society. We in North America, to the contrary, think that it is a quite revolting and totally immoral practice. We do not think that it is something which any dutiful wife should have done to her. We think that it is a way of unfairly keeping women in a subservient role.

Clearly, there is nothing particularly objective about this kind of morality, nor is it something that one would expect to find the inevitable product of natural selection. Even if one thinks there is an indirect connection (and I could think up a scenario to do with religion or sex roles), no one pretends it is something we would expect to find practiced among our extraterrestrials. (These practices presuppose sexuality. In a moment, I will raise again the possibility of extraterrestrial sexuality.)

I am sure some readers will be upset at the moral relativism apparently endorsed in the last two paragraphs (Taylor 1978). It will be argued that suttee and female circumcision are objectively wrong, even though they have been widely endorsed and practiced. They are not – and never could be – subjectively right. I have a sympathy with this objection. My point is that people in different societies have taken different things to be right and wrong. Often, although not necessarily, these relative rights and wrongs have been things bound up with religious practices. I think, for instance, of food taboos and initiation rites. But if we argue – as I probably would – that, overall, religion and surface morality can have biological value, as with the powers of sensation there is clearly more than one way to survive and reproduce (van den Berghe 1979). Hence, at this level I expect our extraterrestrials to differ from us, whatever the ultimate objectivity of the matter may be.

The deeper level to morality

But now let us turn to the deeper level of morality. This is the level which corresponds, as it were, to the epistemological level of the rules of mathematics and causality: the level which informs and gives structure and meaning to our higher level judgments. Here, I see a great deal more universality and common acceptance of moral norms across *Homo sapiens*.[16] Moreover, as intimated above, at this level there is a much more direct connection between moral capacity and adaptive abilities as fashioned by selection. Hence, I have no reason to doubt that some sort of morality would evolve between our extraterrestrials. After all, they have to get on with each other, just as we have to get on with each other. [This assumes they are social. More on this later.]

I have suggested that the extraterrestrials will recognize $2 + 2 = 4$. Will they recognize that which we would think of as the ultimate moral norms? My suspicion is that, in some way, they will. Two of the greatest and most widely accepted enunciations of the supreme principle of morality are the Greatest Happiness Principle and the Categorical Imperative. The former specifies that one's actions ought to be such as will maximize happiness (Mill 1910). The latter, due to Kant, has a number of formulations, but perhaps the easiest to grasp is that which entreats one to regard one's fellow humans as ends, and not simply as means to one's own gratification (Kant 1959). Either or both of these could find their equivalent on our hypothetical planet elsewhere in the universe.

Isn't this just wild speculation? Yes it is – but not that wild! Take the Greatest Happiness Principle. I do not know in what extraterrestrial happiness will consist. Perhaps it will be wallowing in mud baths; but, presumably, extraterrestrials will want to achieve their kind of happiness as best they can. There are good evolutionary reasons for liking sugar, sex, and sleep (Barash 1977; Symons 1979). There are even good reasons for the enjoyment of mental effort. Consequently, I see no reason why it should not be the case that, as it were, there will be some kind of reciprocal feeling between the extraterrestrials, suggesting that, inasmuch as one maximizes the happiness of others, this is a good thing. This feeling, although perhaps unconscious, will occur primarily because if one has such a feeling, others will have such a feeling in reverse for you.

I must qualify this somewhat optimistic conclusion by noting that down here on earth, it is doubtful that many take the Greatest Happiness Principle literally. I am sure most people think they have greater obligations to their children than to others, and (perhaps) to their countrymen than to foreigners (Mackie 1977). Biology can give fairly ready explanations of this sense of variable obligation (in terms of kin selection and the like), and presumably what can be done for terrestrial morality can likewise be done for extraterrestrial morality (Ruse 1984).

The Categorical Imperative seems to be a philosophical version of the biological notion of reciprocal altruism simply made explicit (Murphy 1982). Reciprocal altruism urges you to care about others and to think of their well-being. What is this but to treat others as ends and not simply to use them? Of course, the Categorical Imperative in itself does not look at underlying causes why one should feel the urge to obey it. In fact, as noted, Kant himself did not think that you should ever do anything for hope of conscious return. But, as we have seen, natural selection does not demand conscious awareness of one's motives. In fact, in morality one might be better off doing things because one thinks they are right. (In this sense also, ethics parallels epistemology. The Darwinian is Humean in thinking that morality does not correspond to an external law. As noted, I would rather not talk of 'deception' here.)

What I argue, therefore, is that if, indeed, natural selection is at work on our extraterrestrials, we might expect to find something akin to the Categorical Imperative having evolved amongst them. Indeed, I would go further than this. My own hunch is that a lot of our moral tensions arise because evolution is not desperately concerned with solving every last moral problem. It gives us a rather broad and crude set of moral norms, which includes both the Greatest Happiness Principle and the Categorical Imperative. Usually these norms overlap; but, sometimes they come apart a little. We are better off having both, despite the occasional tensions which arise. Perhaps therefore, our extraterrestrials will have moral philosophers, just as we do. Perhaps, as ours do, their philosophers will agonize over whether or not one should be a utilitarian, or whether one should be a Kantian, or perhaps something else.[17]

Sexual morality

What about sexual morality? Much human moral thought centers on sexual desires and activities. Will there be a sexual morality for our extraterrestrials? Is rape wrong on Andromeda?

An affirmative answer presupposes that our extraterrestrials will be sexual beings. But, as I have pointed out, our present understanding of the evolution and maintenance of sex is not yet strong enough to predict with any reliability that extraterrestrials will necessarily be sexual beings. However, I would hazard a guess that they will be. Although sex certainly did not appear in order to bring about complexity and intelligence – natural selection is not that far-sighted – it is not easy to see how, without any sexuality, great complexity and intelligence could have occurred. Sexuality means good variations can be gathered together in one organism, much more quickly than could be achieved otherwise. Hence, I predict male and female extraterrestrials. (There is not much point in a third sex, although one could have sterile castes.)

But this does not mean we will have boy E.T.s and girl E.T.s, just as here on earth. Everyone might be hermaphroditic. Or, there might be long phases of asexual reproduction, with very rare cases of sexuality. Perhaps indeed the extraterrestrials have reverted entirely to asexuality. Or, the mates might be minute, as in barnacles. Possibly the intelligent extraterrestrials are all female, with little wart-like objects on their skin, these being males – no more than sperm sacs and penises (Darwin 1851, 1854). If this is the way of extraterrestrial sexuality, then there is little need of sexual morality.

But what if the denizens of our hypothetical planet have sexes of comparable size and intelligence, as with humans and other mammals? What place then for sexual morality? Let us take rape as a test case. I realize there are those who argue that rape is not primarily a sexual act, but really some kind of power trip (Brownmiller 1975). If it is simply this, then I presume that it is covered by the Greatest Happiness Principle or the Categorical Imperative. But I would argue – as I am sure most others would – that whatever else may be involved in rape (and probably power is), sex comes in at some point. Without sexuality, the very concept of rape collapses. Moreover – and this is important – almost universally rape is considered immoral. This holds, whether we are talking about eighteenth-century India, twentieth-century Africa, or twentieth-century North America. There are times when one can get away with rape, without too much fear of retribution – during times of warfare, for instance. But, rape is not considered something which is morally accept-able. Rape is wrong.

Nevertheless, we cannot automatically assume that our extraterres-trials would think rape immoral. Why? Because, although the immorality of rape is a human constant, we cannot thereby assume that it will be a constant for other organisms, including extraterrestrial intelligent organisms. Certainly, if we look elsewhere in the animal world, we see that acts which look very much like rape, occur on a regular basis. For instance, among mallards, males will very frequently dash in and copulate violently with a female which has already paired with another male (Barash 1977). Similar behavior occurs in the mammalian world (Wilson 1975; Dawkins 1976). Furthermore, there are good biological reasons why this sort of behavior frequently occurs. If a male animal is prepared to attempt rape on occasion, then he is more like to reproduce than otherwise.

But, you might object: because rape may be frequent in a species, it does not make it moral. In any case, the evidence we have from this earth, from the only extant intelligent species, is that it does not have to be *that* frequent, and does not have to be considered moral. This is true, but one can certainly think of situations where it could be quite frequent, despite intelligence. Suppose the extraterrestrial females all came into heat, as do the females of most mammalian species. If the males simply could not stop themselves from rushing in and copulating, one would have a lot of rape. And, it could hardly be considered that immoral, since no one would have much choice.[18]

Furthermore, consider one of the major biological reasons why we humans think rape wrong – especially why males think rape wrong. Because humans take so long to mature, males cooperate in child-rearing – unlike most other mammals (Lovejoy 1981). Hence, there are good reasons why human morality is trans-sexual, and why nobody (especially no male) wants some third party leaving his seed around in fertile places. If you have got to spend years raising a child, you prefer it to be your own. However, if extraterrestrial females did all of the child-rearing, unaided, one might simply have moral emptiness when it came to rape. This is not to say the females would not have strategies to mate with the 'best' males (Hrdy 1981). They would![19] (I am not saying we think rape is wrong simply because males might have to raise the children of others. I am looking for biological reasons why we might feel so strongly about it. Why is non-physically injurious rape put on a par with assault, or murder even? Do not say 'Because people get upset.' That's the whole point!)

The moral code of the extraterrestrial

Let's sum up. Extraterrestrial ethics will follow extraterrestrial epistemology. At the surface and even intermediate level, we shall find significant differences between extraterrestrials and humans. In fact, the differences may be so great that it is impossible for us to find any common ground. We would simply be dealing with chalk and cheese. If what I have just been saying about sexual morality is correct, then, as with epistemology, these differences might run a lot deeper than most people imagine. We all know of differences between humans, but our common humanity makes for many similarities. We may differ on meat-eating. We agree on rape. Outside of our species the differences multiply, and we may not even find agreement about the morality of rape.

Nevertheless, at the most basic level, there could well be certain fundamental moral similarities between extraterrestrials and humans: that is, assuming some kind of social life occurs. If, like many male mammals, the males lead solitary lives, even this might not be true of males most of the time. But, perhaps naively, I suspect that there is probably either a cause or an effect relationship between intelligence and sociality. Certainly, in our world the two seem to be linked (Lovejoy 1981).

As in the case of epistemology, I have to admit that we are dealing with the products of evolution, and I, myself, am a product of evolution. Could there be a morality quite beyond my comprehension? All I can say is that, as with mathematics and science, I cannot conceive of what a non-human morality, at the most basic level, would be like. Could one have a morality which does not value happiness or which does not entreat one to treat one's fellow species' members as ends in themselves? I cannot conceive of what it would be like. As before, that does not mean that such a morality is, in any sense, likely. It is certainly not something in which it is reasonable to believe.

Hands across the universe?

Finally, what about the question of how extraterrestrials would behave towards us, and how we should behave towards them? The trouble is that neither our sense of morality nor their sense of morality has evolved to deal directly with situations like this. Therefore, in some ultimate sense, I do not think there are any definite answers.

But this is no real cause for concern. I do not know that our sense of morality has really evolved to give us any direct answers about our behavior towards animals. This is surely reflected by the fact that people have different opinions about the proper way to behave towards animals.

And this difference is not simply a question of not having enough information. Some people's intuitions lead them one way, other people's intuitions lead them another way. My intuitions, for instance, lead me to the belief that it is perfectly acceptable to kill animals in order to eat them. Peter Singer, who is a no less rational person than I, has intuitions which lead him to the belief that it is immoral to kill animals in order to eat them (Singer 1975).

Much of the breakdown comes about because our sense of morality has really only evolved in order to enable us to deal with fellow humans. Indeed, I am not even really sure that it is evolved to give us a sense of how we should behave towards all humans. We certainly pay lip service to our need to respect members of remote and foreign cultures as ends in themselves; but, whether we really truly believe that we have a moral obligation to them, is, I think, a very debatable point, as more than one philosopher has pointed out in the past. Remember the qualifications I made earlier when talking of the Greatest Happiness Principle (Mackie 1977; Ruse 1984).

However, although we do not have a well-focussed sense of morality about animals, we are not left completely in the dark. It is not irrational for Singer to try to talk me over, or vice versa. We work by analogy from our feelings about our fellow humans. Much depends on the extent to which we think animals are like us and, also, on the extent to which we relate to animals and they relate back to us. For instance, we tend to treat as moral beings those animals with which we have a close warm relationship, and which respond. We consider their happiness as well as our own; consider them as ends, not simply as means for us. For instance, I would certainly not think of hurting or eating my family dog. I care about his happiness, for his sake.

On the other hand, animals with which I do not have any particular relationship, or animals that do not strike me as particularly intelligent or responsive, receive far less moral concern from me. Although I do not think I should behave in a cruel manner towards pigs, I have few qualms about killing and eating them. I do not relate to pigs as I do to dogs. Moreover, I do not think they suffer a sense of angst at the thought of being killed. So in this sense, also, I do not think I have the obligations to pigs that I have to humans. (I am referring now to pigs' overall lives, and not necessarily to the cruelty which is practiced in many modern slaughterhouses.)[20]

In short, what I am suggesting is that, inasmuch as other animals are like humans, and relate to humans in a reciprocal way, then human morality extends to them. And, inasmuch as animals are not like humans, human morality fades away. Certainly, when dealing with animals which do not respond in a friendly way to me, I do not feel a particular urge to respond in a friendly way to them. I do not, for instance, feel any sense of morality towards snakes; particularly towards those poisonous snakes which would kill me, if it suited their purposes. Nor do I see why I should treat snakes as moral beings. (I would accept qualifications about not being wantonly cruel to snakes or not gratuitously making them go extinct. But why do we preserve such species – for their happiness, or for ours?)

Extraterrestrials merit conclusions analogous to those we draw about non-human terrestrials. Inasmuch as we can relate to them, and they seem to be like us, then no doubt our sense of morality will extend to them. Inasmuch as there is a failure of empathy, then our moral feelings will be that much diminished. Conversely, if, in some way, we can relate to and communicate with the extraterrestrials and they with us, then presumably they will be that much more inclined to treat us as they would their fellow extraterrestrials. The bottom line is their (our) attitude and behavior towards us (them). If they (we) are unbrokenly hostile, or if they (we) seem as though they (we) are going to manipulate us (them) for their (our) ends, then we (they) are going to feel no strong sense of moral obligation towards them (us). Nor indeed, is it easy to see why we (they) should feel any sense of obligation.

To illustrate the point I am making let us pick up where we came into this essay. In the movie E.T., the little boy Eliott, who met the extra-terrestrial, clearly felt a sense of warmth towards E.T., and felt moral sentiments directed towards E.T.'s welfare. Why was this? Simply because E.T., in return, showed friendship and kindliness towards Eliott. And we would agree that it is appropriate to use moral language at this point. Eliott's moral feelings towards E.T. were not misplaced, nor were E.T.'s moral feelings towards Eliott misplaced. It was right for Eliott to help E.T. and vice versa.

However, when we turn to less friendly extraterrestrials, like Darth Vader, the villian in *Star Wars*, it is hard to see why anyone should have any sense of moral obligation to him. No one likes him very much, even his allies, and certainly the heroes of *Star Wars* do not feel any obligations to Darth Vader. Rather, they try to do him down at any and every opportunity. Nor were they wrong in so trying. Darth Vader simply wants to do harm to others; consequently, in return, others feel no obligation towards him.

I would argue that the heroes, in fact, have no real obligation to Vader. He is outside the moral realm, at least as far as Luke Skywalker and friends are concerned. It is interesting to note that Vader's moral reclamation, occurring at the end of *Return of the Jedi*, comes only after Luke sees Vader as his father, recognizes a fellow (albeit fallen) Jedi, and shows love towards him. Vader then, in turn, responds to this love. Luke is his son! A neat case of kin selection at work, possible only when Luke and Vader see analogous beings in each other. Confirmation of my whole line of argument![21]

So, in answer to the question of whether or not we have any moral obligations to extraterrestrials, and whether they have moral obligations to us, all we can say is that 'it all depends.' Are they enough like us that any kind of moral discourse is possible? Is their nature such that they could respond in any kind of friendly way towards us? Is our nature such that it is possible for us to respond in any kind of friendly way to them? There is hardly any point in talking about morality, if, for some reason, the extraterrestrials fill us with such a sense of loathing and disgust and fear that we simply cannot relate to them – and to do so would be deleterious to us. If the possibility of some sort of reciprocal altruism is there, then I see morality emerging; otherwise it may not.

Note: As in the case of terrestrial non-humans, I am trying hard not to slip into some sort of naturalistic fallacy, arguing that morality is simply a function of evolution and that which has evolved is that which is good. I am not endorsing feelings of warmth or hatred, simply because they evolved. I am rather pointing to the fact that morality depends on some kind of moral sense. If a moral sense has not evolved, then morality is simply not possible. Moreover, if moral senses are far out of focus or otherwise blocked, say by mutual animal repulsion or hostility, then morality simply is not going to be able to function. There is no possibility of a moral relationship with extraterrestrials, unless in some way we can both get on the same plane. Perhaps such a plane will exist. As an evolutionist, I do not think it inevitable.

Conclusion
Let me draw the threads together. Exploring the possibility of life elsewhere in the universe is full of philosophical interest. Such exploration puts in a bright light our own powers and limitations. By speculating on what other forms of life would be, we see more clearly the nature and extent of our own knowledge. This is true both in the area of epistemology and in the area of ethics. Such fairy-story telling does not prove anything empirical we do not already know, but it does force us to think again about ourselves from a novel perspective.

In particular, we see a subjective or relative area in both epistemology and ethics. Much that we claim to know, both about the world and about ourselves and our moral relations to others, is very much a function of the kinds of beings we are. Specifically, it is a function of the kinds of animals that we are. We are not possessed of some kind of objective searchlight which enables us to view unfiltered reality. Rather, our vision of the world and our moral feelings for others are direct consequences of our evolutionary past, and there is no further justification.

Yet for all this, there is an underlying 'objective' universality to our knowledge, both in epistemology and in ethics. We may be mere animals, the products of natural selection, but our evolutionary past could not afford to let us believe absolutely anything. We had to be able to respond sensitively and productively to the world out there, and harmoniously with our fellows. The basic principles informing our thinking – principles to which I claim any intelligent (social) being would be sensitive – include such philosophical familiar friends as the central claims of mathematics (in epistemology) and the Greatest Happiness Principle (in ethics). I do not say these principles reflect ultimate reality. Frankly, I do not know what that would mean. I do say that their denial is inconceivable. And that is enough for me.

If the main themes of this essay are correct, that is enough for the inhabitants of Andromeda, also. But, for once, I doubt if anyone is going to show me to be wrong!

Notes

1. By the time you read this, *Return of the Jedi* may be the biggest money-maker. Extraterrestrials abound!
2. For Aristotle, see *On the Generation of Animals*, 561 b and *On the Movement of Animals* 669 b 19. For Kant, see *Cosmologische Briefe* (1761), especially letters 6, 8, and 9.
3. To the second and succeeding editions of his work, Whewell added an ever-growing introduction, responding to critics. His true fears appear even more blatantly in this introduction than in the formal text. In fact, Whewell wrote and had printed an earlier version of the *Plurality*, which included later-excised chapters making his theological motivations more obvious. See Ruse (1975).
4. In a critique of Puccetti (1968), McMullin (1980) makes this point very well. Incidentally, these two contributions to the extraterrestrial life debate show full well that the theological underpinnings of the debate are still with many of us.
5. For a philosopher, I intend to plunge fairly deeply into scientific waters. I will try not to presuppose technical knowledge, but a grasp of modern thought about evolution would help. As background reading let me recommend Dobzhansky *et al.* (1977), Ayala & Valentine (1979), and Futuyma (1979). I give a thoroughly opinionated survey in Ruse (1982*a*). The September 1978 issue of *Scientific American* contains absolutely first-class articles on evolutionary thinking today.

6. Before Darwin, the natural creation of new life – 'spontaneous generation' – was a much discussed topic. However, the canny author of the *Origin of Species* (1859) realized full well that any discussion of the topic would be bound to be incomplete and thus troublesome for his theory. Hence, wisely, Darwin said nothing at all about life's initial creation. And for nearly one hundred years, evolutionists took his advice to heart and likewise said nothing. Only recently, as progress on the coming of life has occurred, has it reappeared in evolutionary discussions. See Farley (1977) for more details.

7. See Popper (1974) for the claim that selection is a tautology, and Ruse (1981*d*) for the denial.

8. For a fascinating discussion of how selection can lead to different strategies for survival, see Oster & Wilson (1978).

9. From here on, I am going to be assuming the general correctness in approach of modern human 'sociobiology,' the neo-Darwinian thesis that human behavioral and intellectual traits and capacities are in major part direct functions of evolution through selection. This was Darwin's own claim in the *Descent of Man* (1871), and it has been taken up recently by such thinkers as E. O. Wilson (1975, 1978); Lumsden & Wilson (1981, 1983); Alexander (1979); Symons (1979); Hrdy (1981); van den Berghe (1979); and Barash (1977). I myself give fairly sympathetic accounts in Ruse (1979*b*, 1981*c*, 1982*a*). I must warn, however, that there are strong and strident critics, for instance Sahlins (1976); and Allen *et al.* (1977). Let me simply say that if the extreme critics are right, and if indeed nothing human (other than our anatomy) has a link to biology, then I just do not know that you could make any claims about extraterrestrial knowledge. But, as I argue in Ruse (1982*a*,*b*), I do not think the critics are right. My arguments in this essay, however, do not depend on some of the more specific and controversial claims by sociobiologists (e.g. Lumsden & Wilson's gene/culture co-evolutionary model).

10. The literature on this subject is so vast that I really do not know where to begin with the references. The major pertinent recent works are probably Griffin (1976), and Crook (1980). A useful short discussion, with many references, is Crook (1983). As you will see in a moment, I rather dodge the whole issue by claiming it is not important (to me)!

11. Indeed, as Kuhn (1962) has truly pointed out, scientists within the same culture frequently find it impossible to communicate, one with another. As will become clear, I do not subscribe to Kuhn's radical position.

12. Putnam invites us to consider the possibility that we are all 'brains in a vat,' that is disembodied brains, supplied with appropriate nutrients and electrical inputs to think that we are normal human beings – seeing, feeling, loving, laughing, and so forth. Such brains have their reality. They 'know' chairs and tables and each other. They are 'internal realists.' But what of the outside world of the vat – 'metaphysical reality'? Putnam argues that such an outside world is an incoherent notion, judged from our human vantage point. Coming from a biological perspective, a perspective on which Putnam has some rather odd ideas, and not having read his work before completing the first draft of this essay, I am nevertheless encouraged to find that I make the same distinctions of reality as he, and that I endorse common-sense (internal) reality, whilst feeling most uncomfortable about ultimate (metaphysical) reality. But we are both Kantians of a kind, so perhaps the consilience was to be expected.

13. This is the famous 'Molyneux problem.' Can a man, blind from birth, able to distinguish cube from sphere by touch, be able to tell cube and sphere as soon as he is given sight? The answer is 'no.' See Berkeley (1963) and especially C. M. Turbayne's introduction.

14. Causation and so forth I take to be governed by kinds of 'epigenetic rules' as discussed by Lumsden & Wilson (1981) – genetically caused traits which affect and govern the way we think. I take them to be akin to the 'regulative principles' of modern Kantians. See Ruse (1982*b*), and Körner (1960).

15. Like all would-be Kantians, I am haunted by the *Ding-an-Sich* problem – how can

one know about ultimate reality? As explained previously, I think that if there is an ultimate reality, a lot of the problems about it are bogus. It does not exist, as it were, like the inside of an orange without the skin (i.e. the external sensations). The inside of an orange is still an object of touch, smell, taste, and so forth. Reality, with sensations stripped away, is not. In fact, 'reality with sensations stripped away' does not make too much sense to me. Putnam (1981) argues that ultimate (metaphysical) reality is really an incoherent notion. Perhaps it is simplest just to leave matters at that.

16. It has always seemed to me that in their desire to hammer E. O. Wilson for being sexist and fascist and so forth, his critics miss entirely the extent to which his (and other sociobiologists') vision of humankind is one of a species *sharing* traits. The New York executive and the Kalahari bushman do what they do for the same reasons. See Ruse (1979*b*) and (1982*a*) for more on this theme.

17. I am not suggesting that all moral disagreements come from clashes of ultimate principles. I am sure that much of the disagreement over (say) censorship comes because we just do not know if pornography corrupts. But, I do suggest that some disagreements come simply because ethics 'break down.' Making an example of a drunk driver has clear utilitarian value – yet, as a Kantian, it just does not seem very fair. Philosophers make a living from arguing their way out of these moral swamps. I wonder if there really is a way out. As an evolutionist rather than Platonist or Christian, knowing that the products of evolution tend to be a bit rough and ready, I am not sure there has to be.

18. One might argue that, by definition, rape occurs only when there is choice, and hence is necessarily wrong. I find this counter unimpressive. I am talking of forcible sexual intercourse, which the male would enjoy, the female would dislike, and which others would not want to see happen. That is close enough to 'rape' for me.

19. Again, let me reassure the reader who fears that sociobiology is all sexist claptrap, that no one is saying that females are defenceless. Unless a female's reproductive strategy is as good as that of a male, she might just as well be one. If a male has (on average) more offspring than a female, then at once there will be strong selective pressures towards giving birth to males. Before long, the rarity of females will make it advantageous to have female offspring, and so the sex–birth ratios will balance out. Sociobiology preaches the biological equality of the sexes.

20. Some readers may think I am cutting awfully close to the naturalistic fallacy here, arguing that because I do not feel friendly towards pigs, I have no moral obligations towards them. This is not so. I certainly do not think I should be wantonly cruel to pigs; but that is because such cruelty lessens me and other humans. It is not because I treat pigs as ends in themselves, or care about pig happiness for its own sake. My point is that you cannot have a moral relationship with something you do not recognize as a moral being. You cannot have a moral relationship with a table. What I am arguing is that moral feelings are a function of evolution, and there is nothing in evolution to say that humans have developed moral feelings for pigs. Indeed, today, evolution emphasizes that selection works ultimately for the individual, not the group. Hence, intra-specific feelings (e.g. those that we humans have for our fellows) have to be related to the individual. Thus my position. Animals become moral beings only insofar as we can extend human-like feelings towards them. I love my dog like a son – literally! See Williams (1966); Dawkins (1976); Ruse (1980*c*).

21. Note also the neat resolution of the Han Solo/Luke Skywalker/Princess Leia triangle. In *The Empire Strikes Back* one starts to feel all sorts of tensions as Han and Leia draw together, apparently excluding Luke. But then it turns out in *Return of the Jedi* that Luke and Leia are siblings. Down comes the incest barrier! This is one of the big moral absolutes for sociobiologists, given the horrendous biological effects of close inbreeding (Alexander 1979; Wilson 1978; van den Berghe 1979). Obviously natural selection has been at work in George Lucas' world, just as I predicted it would!

References

Alexander, R. D. (1971). The search for an evolutionary philosophy. *Proceedings of the Royal Society of Victoria*, **84**, 99–120.
- (1974). The evolution of social behavior. *Annual Review of Ecology and Systematics*, **5**, 325–84.
- (1975). The search for a general theory of behavior. *Behavioral Science*, **20**, 77–100.
- (1979). *Darwinism and Human Affairs*. Seattle: Washington University Press.
Allen, E. *et al.* (1977). Sociobiology: a new biological determinism. In *Biology as a Social Weapon*, ed. Sociobiology Study Group of Boston, pp. 133–49 Minneapolis: Burgess.
Ayala, F. J. & Valentine, J. W. (1979). *Evolving: The Theory and Processes of Organic Evolution*. California: Benjamin, Cummings.
Barash, D. P. (1977). *Sociobiology and Behavior*. New York: Elsevier.
Beck, L. W. (1972). Extraterrestrial intelligent life. *Proceedings and Addresses of the American Philosophical Association*, XLV, 5–21.
Berkeley, G. (1963). *Works on Vision*. Indianapolis: Library of Liberal Arts.
Billingham, J. (1981). *Life in the Universe*. Cambridge, Mass.: MIT Press.
Black, M. (1962). *Models and Metaphors*. Ithaca, NY: Cornell University Press.
Brewster, D. (1854). *More Worlds than One: The Creed of the Philosopher and the Hope of the Christian*. London: Murray.
Brooke, J. H. (1977). Natural theology and the plurality of worlds: Observations on the Brewster–Whewell debate. *Annals of Science*, **34**, 221–86.
Brownmiller, S. (1975). *Against Our Will: Men, Women, and Rape*. New York: Simon and Schuster.
Chomsky, N. (1957). *Syntactic Structures*. The Hague: Mouton.
Crook, J. H. (1980). *The Evolution of Human Consciousness*. Oxford: Oxford University Press.
- (1983). On attributing consciousness to animals. *Nature*, **303**, 11–14.
Darwin, C. (1851). *A Monograph of the Sub-class Cirripedia, with Figures of all the Species, The Lepadidae; or Pedunculated Cirripedes*. London: Royal Society.
- (1854). *A Monograph of the Sub-class Cirripedia, with Figures of all the Species. The Balanidae (or Sessile Cirripedes); the Verrucidae, etc.* London: Ray Society.
- (1859). *On the Origin of Species*. London: John Murray.
- (1871). *Descent of Man*. London: Murray.
Dawkins, R. (1976). *The Selfish Gene*. Oxford: Oxford University Press.
de Waal, F. (1982). *Chimpanzee Politics: Power and Sex Among Apes*. New York: Harper and Row.
Dick, S. (1982). *Plurality of Worlds: the Origins of the Extraterrestrial Life Debate from Democritus to Kant*. Cambridge: Cambridge University Press.
Dickerson, R. E. (1978). Chemical evolution and the origin of life. *Scientific American*, **239** (3), 70–86.
Dobzhansky, Th. (1951). *Genetics and the Origin of Species*, 3rd edn. New York: Columbia University Press.
- (1962). *Mankind Evolving*. New Haven: Yale University Press.
- (1970). *Genetics of the Evolutionary Process*. New York: Columbia.
Dobzhansky, Th., Ayala, F. J., Stebbins, G. L., & Valentine, J. W. (1977). *Evolution*. San Francisco: Freeman.

Eigen, M., Gardiner, W., Schuster, P., & Winkler-Oswatitsch, R. (1981). The origin of genetic information, *Scientific American*, **224** (4), 88–118.

Endler, J. A. (1977). *Geographic Variation, Speciation, and Clines*. Princeton: Princeton University Press.

Farley, J. (1977). *The Spontaneous Generation Controversy: From Descartes to Oparin*. Baltimore: Johns Hopkins Press.

Feduccia, A. (1980). *The Age of Birds*. Cambridge, Mass.: Harvard University Press.

Flew, A. G. N. (1967). *Evolutionary Ethics*. London: Macmillan.

Futuyma, D. (1979). *Evolutionary Biology*. Sunderland, Mass.: Sinauer.

Ghiselin, M. (1974). *The Economy of Nature and the Evolution of Sex*. Berkeley: University of California Press.

Grant, V. (1971). *Plant Speciation*. New York: Columbia University Press.

– (1981). The genetic goal of speciation. *Biologisches Zentralblatt*, **100**, 473–82.

Griffin, D. A. (1976). *The Question of Animal Awareness: Evolutionary Continuity of Mental Experience*. New York: Rockefeller University Press.

Grossman, N. (1974). Empiricism and the possibility of encountering intelligent beings with different sense-structures. *Journal of Philosophy*, LXXI, 815–21.

Hamilton, W. D. (1964a). The genetical evolution of social behaviour. I. *Journal of Theoretical Biology*, 7, 1–16.

– (1964b). The genetical evolution of social behaviour. II. *Journal of Theoretical Biology*, 7, 17–32.

Hart, M. H. & Zuckerman, B. (eds.) (1982). *Extraterrestrials: Where Are They?* New York: Pergamon.

Hesse, M. (1966). *Models and Analogies in Science*. Notre Dame, Ind.: University of Notre Dame Press.

Hrdy, S. (1981). *The Woman that Never Evolved*. Cambridge, Mass.: Harvard University Press.

Huxley, T. H. (1901). *Evolution and Ethics, and Other Essays*. London: Macmillan.

Johanson, D. & Edey, M. (1981). *Lucy: The Beginnings of Humankind*. New York: Simon and Schuster.

Johanson, D. & White, T. D. (1979). A systematic assessment of early African hominids. *Science*, **203**, 321–30.

Kant, I. (1959). *Foundations of the Metaphysics of Morals*. Trans. L. W. Beck. Indianapolis: Bobbs-Merrill.

Körner, S. (1960). On philosophical arguments in physics. In *The Structure of Scientific Thought*, ed. E. H. Madden, pp. 106–10. Boston: Houghton, Mifflin.

Kuhn, T. S. (1962). *The Structure of Scientific Revolutions*. Chicago: University of Chicago Press.

Laudan, L. (1977). *Progress and its Problems: Towards a Theory of Scientific Growth*. Berkeley: University of California Press.

Levin, D. A. (1979). The nature of plant species. *Science*, **204**, 381–4.

Lewontin, R. C. (1978). Adaptation. *Scientific American*, **239** (3), 212–30.

Lovejoy, C. O. (1981). The origin of man. *Science*, **211**, 341–50.

Lumsden, C. J. & Wilson, E. O. (1981). *Genes, Mind and Culture: The Coevolutionary Process*. Cambridge, Mass.: Harvard University Press.

– (1983). *Promethean Fire*. Cambridge, Mass.: Harvard University Press.

Mackie, J. (1977). *Ethics: Inventing Right and Wrong*. Harmondsworth, Mddx.: Penguin.

McMullin, E. (1980). Persons in the universe. *Zygon*, **15**, 69–89.

Maynard Smith, J. (1978). The evolution of behavior. *Scientific American*, **239** (3), 176–193.

‒ (1982). *The Evolution of Sex.* Cambridge: Cambridge University Press.
Mayr, E. (ed.) (1957). *The Species Problem.* AAAS Pub. No. 50. Washington, DC.
Mayr, E. (1963). *Animal Species and Evolution.* Cambridge, Mass.: Belknap.
‒ (1982). *The Growth of Biological Thought: Diversity, Evolution, and Inheritance.* Cambridge, Mass.: Harvard University Press.
Mill, J. S. (1910). *Utilitarianism, Liberty, and Representative Government.* London: Dent.
Murphy, J. G. (1982). *Evolution, Morality, and the Meaning of Life.* Totowa, N.J.: Rowman and Littlefield.
Oster, G. F. & Wilson, E. O. (1978). *Caste and Ecology in the Social Insects.* Princeton: Princeton University Press.
Popper, K. (1974). Darwinism as a metaphysical research programme. In *The Philosophy of Karl Popper*, ed. P. A. Schilpp, pp. 133–43. LaSalle, Ill.: Open Court.
Puccetti, R. (1968). *Persons: A Study of Possible Moral Agents in the Universe.* London: Macmillan.
Putnam, H. (1981). *Reason, Truth, and History.* Cambridge: Cambridge University Press.
Quine, W. V. O. (1969). Natural kinds. In *Ontological Relativity and Other Essays*, pp. 114–38. New York: Columbia University Press.
Raup, D. M. & Stanley, S. M. (1978). *Principles of Paleontology*, 2nd edn. San Francisco: Freeman.
Ruse, M. (1973*a*). The value of analogical models in science. *Dialogue*, **12**, 246–53.
‒ (1973*b*). The nature of scientific models: formal *v.* material analogy. *Philosophy of the Social Sciences*, **3**, 63–80.
‒ (1975). The relationship between science and religion in Britain, 1830–1870. *Church History*, **44**, 505–22.
‒ (1977). Is biology different from physics? In *Logic, Laws, and Life*, ed. R. Colodny, pp. 89–127. Pittsburgh: University of Pittsburgh Press.
‒ (1979*a*). *The Darwinian Revolution: Science Red in Tooth and Claw.* Chicago: University of Chicago Press.
‒ (1979*b*). *Sociobiology: Sense or Nonsense?* Dordrecht: Reidel.
‒ (1980*a*). Philosophical aspects of the Darwinian Revolution. In *Pragmatism and Purpose*, ed. F. Wilson, pp. 220–35. Toronto: University of Toronto Press.
‒ (1980*b*). Ought philosophers consider scientific discovery? A Darwinian case study. In *Scientific Discovery*, ed. T. Nickles, pp. 131–49. Dordrecht: Reidel.
‒ (1980*c*) Charles Darwin and group selection. *Annals of Science*, **37**, 615–30.
‒ (1981*a*). Teleology redux. In *Essays in Honour of Mario Bunge*, ed. J. Agassi, pp. 299–309. Dordrecht: Reidel.
‒ (1981*b*). *Is Science Sexist? And Other Problems in the Biomedical Sciences.* Dordrecht: Reidel.
‒ (1981*c*). Is human sociobiology a new paradigm? *Philosophical Forum*, **13**, 119–43.
‒ (1981*d*). Darwin's theory: an exercise in science. *New Scientist*, **90**, 828–30.
‒ (1982*a*). *Darwinism Defended: A Guide to the Evolution Controversies.* Reading, Mass.: Addison-Wesley.
‒ (1982*b*). Darwin's legacy. In *Charles Darwin: A Centennial Commemorative*, ed. R. Chapman, pp. 323–54. Wellington, NZ: Nova Pacifica.
‒ (1984). The morality of the gene. *The Monist*, **67** (2), 167–99.
Sahlins, M. (1976). *The Use and Abuse of Biology.* Ann Arbor: University of Michigan Press.

Schopf, J. W. (1978). The evolution of the earliest cells. *Scientific American*, **239** (3), 110–38.

Singer, P. (1975). *Animal Liberation*. New York: New York Review.

– (1981). *The Expanding Circle: Ethics and Sociobiology*. New York: Farrar, Straus, and Giroux.

Sokal, R. R. (1973). The species problem reconsidered. *Systematic Zoology*, **22**, 360–74.

Stebbins, G. L. & Ayala, F. J. (1981). Is a new evolutionary synthesis necessary? *Science*, **213**, 967–71.

Symons, D. (1979). *The Evolution of Human Sexuality*. New York: Oxford University Press.

Taylor, P. W. (1978). *Problems of Moral Philosophy*. Belmont, Calif.: Wadsworth.

Trivers, R. L. (1971). The evolution of reciprocal altruism. *Quarterly Review of Biology*, **46**, 35–57.

– (1972). Parental investment and sexual selection. In *Sexual Selection and the Descent of Man, 1871–1971*, ed. B. Campbell, pp. 136–79. Chicago: Aldine.

– (1976). 'Foreword' to R. Dawkins, *The Selfish Gene*, pp. v–vii. Oxford: Oxford University Press.

– (1983). The evolution of sex. *Quarterly Review of Biology*, **58**, 62–7.

Valentine, J. W. (1978). The evolution of multicellular plants and animals. *Scientific American*, **239** (3), 140–58.

van den Berghe, P. L. (1979). *Human Family Systems: An Evolutionary View*. New York: Elsevier.

Van Valen, L. (1976). Ecological species, multispecies, and oaks. *Taxon*, **25**, 233–9.

Whewell, W. (1853). *Of the Plurality of Worlds: An Essay*. London: Parker.

White, M. J. D. (1978). *Modes of Speciation*. San Francisco: Freeman.

Williams, G. C. (1966). *Adaptation and Natural Selection: A Critique of Some Current Evolutionary Thought*. Princeton: Princeton University Press.

– (1975). *Sex and Evolution*. Princeton: Princeton University Press.

Wilson, E. O. (1975). *Sociobiology: The New Synthesis*. Cambridge, Mass.: Harvard University Press.

– (1978). *On Human Nature*. Cambridge, Mass.: Harvard University Press.

PART 3

Extraterrestrial epistemology

Although Raup has suggested the possibility of our receiving electromagnetic radiations from non-intelligent aliens who involuntarily broadcast signals into space, the usual expectation is that we will receive messages that have been intentionally formulated, encoded, and transmitted by the others. These messages need not have been intended for our consumption. Indeed, the first alien transmissions that we might detect could well be leakages of radio signals from another planetary surface. Our own radio signals have been leaking out into space for the last 50 years, and another species – so long as they used powerful enough reception apparatus that was pointed in the right direction – could have intercepted some of these signals.

But what could a hypothetical alien learn about intelligent life on earth from our electromagnetic seepages into space? The answer is: a great deal. W. T. Sullivan, S. Brown, and C. Wetherill (in *Science* (1978), **199**, 377–88) analyzed what aliens would be able to learn about us if they had a receiver as large as the Cyclops listening system (an 8 km array of 1500 radio dishes) which SETI scientists have proposed be built on earth. Even if they were as far as 25 light years away, they would be able to deduce the existence of a technology on earth, and even be able to draw a map showing where that technology is concentrated. From the rising and setting times of earth's radio stations the aliens could deduce earth's rotational velocity and size, and the latitude of its sending antennas. From observations over a one-year period an alien radioastronomer could find the shape, duration, and size of earth's orbit around the sun. From detailed analysis of incoming signals, the aliens might even be able to deduce the size of our sending antennas, and hence make inferences as to the size of artificial structures on earth. There's of course no reason at all to think that this has ever happened, but at least in principle it seems that alien cultures could have learned these things about us already.

If others could learn such things about us in these ways, then we could learn such things about them. Even in the absence of intentional broadcasts for our benefit, we could learn things about other technical civilizations from the clues contained in their involuntary emissions into space. Totally unawares, they may be sending information to us that we would be able to detect and interpret, on the basis of which we could conclude that intelligent extraterrestrial life does exist.

More commonly, however, SETI theorists posit that aliens will send messages to other cultures intentionally, just as we on earth have already done. We have sent messages in physical media on spacecraft – the Pioneer plaques and the Voyager Interstellar Record – and a single electromagnetic message – the Arecibo Interstellar Message, November 1974. (For details, see Carl Sagan *et al.*, *Murmurs of Earth*, New York, Ballantine, 1978.)

Plainly, if an alien culture is to send such messages to us, they must have intelligence, science, and technology. Further, if we are to understand those messages, their intelligence, science, and technology must be compatible with our own. In this section, two authors present contrasting views on whether such compatibility is likely to exist. If it does not then our listening programs will be doomed from the start.

Nicholas Rescher, Professor of Philosophy at the University of Pittsburgh, and author of several books on philosophy and science, argues that extraterrestrials are extremely unlikely to have any type of science that would be recognizable to us. The reasons for this are manifold, but paramount among them are that (1) they will almost certainly be different organisms, with different needs, senses, and behaviors; (2) they will inhabit environments strikingly different from our own, environments in which neither science nor technology may be needed for survival; (3) since several species on earth have survived longer than man, intelligence is not a necessary adaptation, and should not be expected to develop elsewhere.

Further, even if an alien culture does have some kind of science, Rescher argues that it probably will not be one that we could understand. The development of science even on earth shows that the science of one era may be totally incomprehensible to that of another. The two do not even talk about the same things: 'It's not that modern medicine has a different theory of the operation of the four humors than Galenic medicine,' says Rescher, 'but that modern medicine has *abandoned* the four humors. It's not that the Galenic physician says different things about bacteria and viruses, but that he has *nothing* to say about them.' The acceleration of change in scientific theory in recent times only heightens the problem. Earthly science in 100 years would be unintelligible to us today.

. If the type of sea change that Rescher describes characterizes earthly science over a very short period of time in cosmological terms, there is virtually no hope that the sciences of two galactic cultures separated by a chasm of space and time will be in any way comparable with one another. Whereas Raup in the previous section held out the chance of a non-technical species sending electromagnetic waves that we could recognize as coming from extraterrestrial life, Rescher presents a picture of an intelligent, perhaps even technological and scientific species whose existence will be forever undetectable by us.

A different view is advanced by MIT's artificial intelligence pioneer Marvin Minsky. (An extended profile of Minsky appears in Jeremy Bernstein's *Science Observed*, New York, Basic Books, 1982.) Drawing upon AI (Artificial Intelligence) researches into what intelligence – whether electronic or natural – is, and upon general problem-solving strategies, Minsky claims to explain 'Why intelligent aliens will be intelligible.'

Contrary to Rescher, Minsky argues that intelligent extraterrestrials 'will think like us, in spite of different origins.' Minsky reasons that all intelligent problem solvers are subject to the same ultimate constraints: limitations on space, time, and materials. Two principles, of Sparseness and of Economics, show that every intelligence will be forced to develop an arithmetic, and a language whose structures are rooted in the natures of things. Because arithmetic is the same everywhere, alien mathematics necessarily will be congruent to our own. And because things are, in their most general aspects, the same everywhere, aliens will have evolved thought-processes and languages that will match our own to a degree that will enable us to comprehend them.

Minsky responds in advance to the countercharges of possible critics. To the suggestion, for example, that aliens might know things holistically instead of by the discrete and separate thought-processes of the human mind, Minsky replies that holistic ways of knowing would surrender the very thing that makes any complex problem soluble: its being broken down into parts. There is only one way for intelligence to deal with reality, Minsky suggests, and we and aliens will have this in common. That is why we will be able to communicate with them.

Extraterrestrial science

NICHOLAS RESCHER

Could science in another setting overcome the limitations of our human science?

It is tempting to wonder whether another, astronomically remote civilization might be scientifically more advanced than ourselves. Is it not plausible to suppose that an alien civilization might improve on our performance in science and manage to surpass us in the furtherance of this enterprise? The question seems very clear-cut on first thought, for, as one recent discussion puts it:

> 'Any serious speculations concerning the capabilities of intelligent biological life and automata must take into account technical societies that may be millions or even billions of years more advanced than our own.'
> (MacGowan & Ordway, 1966, p. 248.)

This seemingly straightforward issue is one of enormous complexity. And this complexity relates not only to the actual or possible facts of the matter but also – crucially – to somewhat abstruse questions about the very ideas or concepts that are relevant here.

To begin with, there is the question of just what it means for there to be another 'science-possessing' civilization. Note that this is a question that *we* are putting – a question posed in terms of the applicability of *our* term 'science.' It pivots on the issue of whether *we* would be prepared to call certain of *their* activities – once we understood them – as engaging in scientific inquiry and whether *we* would be prepared to recognize the product of these activities as constituting a (state of a branch of) science.

At the very least this requires that we are prepared to recognize what those aliens are doing as a matter of forming beliefs (theories) as to how things work in the world and to acknowledge that they are involved in testing these beliefs observationally or experimentally and in applying them in practical (technological) contexts. We must, to begin with, be prepared to accept them as persons of some sort, duly equipped with intellect and will, and must then enter upon a complex series of claims with respect to their beliefs and their purposes.

A scientific civilization is not merely one that possesses intelligence and social organization, but one that puts this intelligence and organization to work in a very particular sort of way. This opens up a rather subtle issue of priority in regard to process v. product, namely whether what counts for a civilization's 'having a science' is primarily a matter of the substantive *content* of their doctrines (their belief structures and theory complexes) or is primarily a matter of the aims and *purposes* with which their doctrines are formed.

The matter of content turns on the issue of how similar their scientific beliefs are to ours. And this is clearly something on which we would be ill advised to put much emphasis at the very outset. After all, the speculations of the nature-theorists of pre-Socratic Greece, our ultimate ancestors in the scientific enterprise, bear little resemblance to our present-day sciences, nor does contemporary physics bear all that much doctrinal resemblance to that of Newton. We would do better to give prime emphasis to matters of process and purpose.

Accordingly, the matter of these aliens 'having a science' is to be regarded as turning not on the extent to which their *findings* resemble ours, but on the extent to which their *project* resembles ours – of deciding whether we are engaged in the same sort of inquiry in terms of the sorts of issues being addressed and the ways in which they are going about addressing them. The issue accordingly is, at bottom, not one of the substantive similarity of their 'science' to ours but one of the functional equivalency of the projects at issue in terms of the characterizing goals of the scientific enterprise: description, explanation, prediction, and control over nature.

The potential diversity of 'science'
The indicated perspective enjoins the pivotal question: To what extent would the *functional equivalent* of natural science built up by the inquiring intelligences of an astronomically remote civilization be bound to resemble our science? In considering this issue, one soon comes to realize that there is an enormous potential for diversity here.

To begin with, the *machinery of formulation* used in expressing their science might be altogether different. Specifically, their mathematics might be very unlike ours. Their 'arithmetic' could be non-numerical – purely comparative, for example, rather than quantitative. Especially if their environment is not amply endowed with solid objects or stable structures – if, for example, they were jellyfish-like creatures swimming about in a soupy sea – their 'geometry' could be something rather strange, largely topological, say, and geared to structures rather than sizes or shapes. Digital thinking in all its forms might be undeveloped, while certain sorts of analog reasoning are highly refined. Or again, an alien civilization might, like the ancient Greeks, have 'Euclidean' geometry without analysis. In any event, given that the mathematical mechanisms at their disposal could be very different from ours, it is clear that their description of nature in mathematical terms could also be very different. (And not necessarily truer or falser, but just different.)

Secondly, the *orientation* of the science of an alien civilization might be very different. All their efforts might conceivably be devoted to the social sciences – to developing highly sophisticated analysis of psychology and sociology, for example. But their approach to natural science might also be very different. Communicating by some sort of 'telepathy' based upon variable odors or otherwise 'exotic' signals, they might devise a complex theory of thought-wave transmission through an ideaferous aether. Electromagnetic phenomena might lie altogether outside the ken of alien life-forms; if their environment does not afford them loadstones and electrical storms, etc., the occasion to develop electromagnetic theory might never arise. The course of scientific development tends to flow in the channel of practical interests. A society of porpoises might lack crystallography but develop a very sophisticated hydrodynamics; one comprising mole-like creatures might never dream of developing optics. One's language and thought-processes are bound to be closely geared to the world one experiences. As the difficulties we experience in bringing the language of everyday experience to bear on subatomic phenomena illustrates, our concepts are ill attuned to facets of nature different in scale or structure from our own. We can hardly expect a 'science' that reflects our parochial preoccupations to be a universal constant. The science of a different civilization would presumably be closely geared to the particular pattern of their interaction with nature, as funnelled through the particular course of their evolutionary adjustment to their specific environment.

Again, alien civilizations might scan nature very differently. The 'forms of sensibility' of radically different life-forms (to invoke Kant's useful idea) are bound to be radically different from ours. The direct chemical analysis of environmental materials might prove highly useful, and bioanalytic techniques akin to our senses of taste and smell could be highly developed and provide the basis for sciences of different sorts. Or think of the beings who would populate Strawson's purely auditory world (Strawson 1959, pp. 59–86). Acoustics might mean very little to them, while other sorts of pressure phenomena – for example, the theory of turbulence in gasses – might be the subject of intense and exhaustive investigation. Rather than sending radiowaves or heat signals, they might propel gravity-waves through space. A comparison of the 'science' of different civilizations here on earth suggests that it is not an outlandish hypothesis to suppose that the very *topics* of their science might differ radically from those of ours. In our own case, for example, the fact that we live on the surface (unlike whales or porpoises), the fact that we have eyes (unlike worms or moles) and thus can *see* the heavens, and the fact that we are so situated that the seasonal positions of heavenly bodies are intricately connected with our biological needs through the agricultural route to food supply, are clearly connected with development of astronomy. Then, too, their mode of emplacement in nature might be so different as to focus on entirely different aspects or constituents of the cosmos. If the world is sufficiently complex they might focus on aspects of it that mean little or nothing to us. As one able physicist has written:

> 'Why should Nature run on just a finite number of different types of force? May there not be an infinite hierarchy of types of force just as there is an infinity of structures that may be built of matter interacting under the influence of any one or more of those forces? . . . So, perhaps, our discovery of a sufficient proliferation of types of force may guide us into an empirical classification of those forces in terms of their relationship to one another – analogous to the grouping of the chemical elements in the Mendeleev table – and then to an understanding of that classification in terms of some underlying principle that would then enable us to predict the nature of undiscovered and undiscoverable forces . . . . It is perfectly possible that there exist objects that interact powerfully with each other but only exceedingly feebly with the objects with which we are familiar, that is to say with objects that interact strongly with ourselves, so that these unknown objects could build up complex structures that could share our natural world but of which we would be ignorant.'
> (Wilkinson, 1977, pp. 13–14.)

Accordingly, the constitution of the alien inquirers – physical, biological, and social – emerges as a crucial element here. It serves to determine the agenda of questions and the instrumentalities for their resolution – to fix what is seen as interesting, important, relevant, significant. In determining what is seen as an appropriate question and what is judged as an admissible solution, the cognitive posture of the inquirers must be expected to play a crucial role in shaping and determining the course of scientific inquiry itself.

Thirdly, the *conceptualization* of their science might be very different. We must reckon with the theoretical possibility that a remote civilization might operate with a radically different system of concepts in its cognitive dealings with nature. It is (or should be) clear that there is no single, unique, ideally adequate concept-framework for 'describing the world.' The botanist, horticulturist, landscape gardener, farmer, and painter will operate from diverse cognitive 'points of view' to describe one selfsame vegetable garden in very different terms of reference. It is mere mythology to think that the 'phenomena of nature' can lend themselves to only one correct style of descriptive and explanatory exposition. There is surely no 'ideal scientific language' that has a privileged status for the characterization of reality. To insist on the ultimate uniqueness of science is to succumb to 'the myth of the God's eye view.' Different cognitive perspectives are possible – no one of which is more adequate or more correct than any other, independently of the aims and purposes of their users.

To motivate this idea of a conceptually different science, it helps to cast the issue in temporal rather than spatial terms. The descriptive characterization of *alien* science is a project rather akin in its degree of difficulty to describing our own *future* science. It is a key fact of life in this domain that ongoing progress in the science/technology area is a progress of *ideational* innovation that always places certain developments outside the intellectual horizons of earlier workers. The very concepts we use become available only in the course of scientific discovery itself. Short of learning our science from the ground up, Aristotle could not have made anything of modern genetics nor Newton of quantum physics. The key discoveries of later stages are those which the workers of a substantially earlier period had not only failed to make but which they could not even have understood, because the conceptual instruments for such understanding were not available to them. It is thus effectively impossible to predict not only the answers but even the questions that lie on the agenda of future science, because these questions will grow out of the answers we obtain at yet unattained stages of the game. And the situation of an alien science could be much the same. As with the science of the remote future, the science of the remotely distant must be presumed to be of such a nature that we really could not achieve intellectual access to it on the basis of our own position in the cognitive scheme of things. Just as the technology of a more advanced civilization would be bound to strike us as magic, so its science would be bound to strike us as incomprehensible gibberish – until we had learned it 'from the ground up.' They might (just barely) be able to *teach* it to us, but they could not *explain* it to us.

The most characteristic and significant sort of difference between one conceptual scheme and another thus arises when the one scheme is committed to something the other does not envisage at all – something that lies outside the conceptual range of the other. The typical case is that of the stance of Cicero's thought-world with regard to questions of quantum electrodynamics. The Romans of classical antiquity did not hold *different* views on these issues: they held no views at all regarding them, because they lay outside their conceptual reach. This whole set of relevant considerations simply lay outside their conceptual repertoire.

The 'science' of different civilizations may well, like Galenic and Pasteurian medicine, simply *change the subject* in key respects so as no longer 'to talk about the same things,' but treat things (e.g. humors and bacteria, respectively) of which the other takes no cognizance at all. The difference in 'conceptual scheme' between modern and Galenic medicine is not that the modern physician has a different theory of the operation of the four humors from his Galenic counterpart, but that modern medicine has *abandoned* the four humors, and not that the Galenic physician says different things about bacteria and viruses, but that he has *nothing* to say about them – that they lie entirely beyond his conceptual horizon. If, for example, certain intelligent aliens should prove a diffuse and complex aggregate mass of units comprising wholes in ways that allow of overlaps (compare Ehrensvärd 1965, pp. 146–8), then the role of social concepts might become so paramount that nature as a whole comes to be viewed in these terms. The result would be something very difficult for us to grasp, seeing that they are based on a mode of 'shared experience' with which we have no contact.

It is only reasonable to presume that the conceptual character of the 'science' of an alien civilization is so radically different from ours as to orient their thought about 'the nature of things' in altogether different directions. Even in strictly functional terms, their 'functional equivalent' of our science might be very unlike ours. Their explanatory mechanisms, their predictive concerns, and their modes of control over nature might all be very different. In all these regards they might have interests that depart significantly from our own. It is only natural to expect that creatures with interests shaped under the remorseless pressure of evolutionary adaptations to fortuitous – and endlessly variable – environmental conditions should be oriented in directions very different from ours.

As long as the fundamental categories of thought – the modes of spatiality and temporality, of structural description and the like – are not seen as necessary features of intelligence as such, but as evolved cognitive adaptations to particular contingently constituted ways of emplacement in and interaction with nature, there will be no basis for claims to uniformity. For the prospects of diversity in these regards are literally endless.

Natural science – broadly construed as inquiry into the ways of nature – is something that is in principle almost infinitely plastic. Its development will trace out a historical course closely geared to the specific capacities, interests, environment, and opportunities of the creatures that develop it. We are deeply mistaken if we think of it as a process that must follow a route roughly parallel to ours and issue in a comparable product. It would be grossly unimaginative to think that either the journey or the destination must be the same – or even substantially similar.

The more deeply one reflects on the matter, the more firmly one is led to the realization that our particular human conception of the issues of science is something parochial, because we are physically, perceptually, and cognitively limited and conditioned by our specific mode of emplacement within nature. Given that we might be dealing with intelligent beings of physical and cognitive nature profoundly different from ours, who would dare to claim with confidence what the natural science of such creatures would be like?

The one-world, one-science argument

One recent writer raises the question 'What can we talk about with our remote friends?', and answers that 'We have a lot in common. We have mathematics in common, and physics, and astronomy.' (E. Purcell in Cameron 1963, p. 142.) Another writer maintains that 'we may fail to enjoy their music, understand their poetry, or approve their ideals; but we can talk about matters of practical and scientific concern.' (Anderson 1963, p. 130.) This line of thought begs some very big questions.

Our alien colleagues scan nature for regularities using (at any rate to begin with) the sensors provided them by their evolutionary heritage. They note, record, and transmit those regularities which they find to be intellectually interesting or pragmatically useful. And they develop their inquiries by theoretical triangulation that proceeds from the lines indicated by these resources. Now this is clearly going to make for a course of development that closely gears their science to their particular situation – their biological endowment ('their sensors'), their cultural heritage ('what is interesting'), their environmental niche ('what is pragmatically useful'). Where these key parameters differ, there too we can expect that the course of scientific development will differ as well.

Admittedly there is only one universe, and its laws, as best we can tell, are everywhere the same. We share the universe with all life forms. However radically we differ in other respects (in particular those relating to environment, to forms of life, and its mode of civilization) we have a common background of cosmic evolution and a common heritage of natural laws. And so if intelligent aliens investigate nature at all, they will investigate the same nature we ourselves do. But the sameness of the object of contemplation does nothing to guarantee the sameness of the ideas about it. It is all too familiar a fact that even where human (and thus *homogeneous*) observers are at issue, different constructions are often placed upon 'the same' occurrences. Primitive peoples thought the sun a god, and the most sophisticated among the ancients thought it a large mass of fire. We think of it as a large thermonuclear reactor, and heaven only knows how our successors will think of it in 3000 A.D. As the course of human history clearly shows, there need be little uniformity in the conceptions held about one selfsame object by different groups of thinkers.

It is surely naive to think that because one selfsame object is in question, its description must issue in one selfsame result. This view ignores the crucial matter of one's intellectual orientation. One selfsame piece of driftwood is viewed very differently indeed by the botanist, the painter, the interior decorator, the chemist, the woodcarver, etc. It is a matter of what 'aspects' of the thing are focused upon as important or interesting. Minds with different sorts of concerns and interests and different backgrounds of information can deal with mutually common items in ways that yield wholly disjoint and disparate results because altogether different features of the thing are being addressed. The *things* are the same, but their *significance* is altogether different. Different cognitive vantage points, different 'perspectives of consideration' are at issue.

Different conceptual frameworks take us into literally different spheres of thought. They go their separate ways in very much the same sort of way as different subject-matter specialties dealing with altogether different sorts of issues even when (as viz. the soldier, the farmer, and the geologist) they address themselves to something which appears as 'the same thing' to a detached bystander (the same piece of terrain). It is notorious that observers are prisoners of their cognitive preparation, interests, and predispositions – seeing only what their pre-established cognitive resources enable them to see and blind to that for which they are mentally unprepared.

Accordingly, the sameness of nature and its laws by no means settles the issue of scientific uniformity. For science is always the result of *inquiry* into nature and this is inevitably a matter of a *transaction* or *interaction* in which nature is but one party and the inquiring beings another. The result of such an interaction depends crucially on the contribution from both sides – from nature and from the intelligences that interact with it. A kind of 'chemistry' is at work where nature provides only one input and the inquirers themselves provide another – one that can massively and dramatically affect the outcome in such a way that we cannot disentangle the respective contributions of nature and the inquirer. Things cannot of themselves dictate the significance an active intelligence can attach to them.

No one who has observed how vastly differently the declarations of a single text – the Bible, say, or the dialogues of Plato – have been interpreted and understood over the centuries, even by people of a shared cultural heritage, can be hopeful that the study of a common object by different civilizations must lead to a uniform result. Yet such analogies are oversimple and misleading, because with scientific inquiry there is no one predetermined original text – no 'God's eye' science. It would be a grave mistake to think that there is just one fixed basic text – the 'book of nature writ large' – which different civilizations can decipher in different degrees.

After all, throughout the earlier stages of man's intellectual history, different human civilizations have developed their 'natural sciences' in substantially different ways. And the shift to an extraterrestrial perspective is bound to amplify such cultural differences. The 'science' of an alien civilization may be far more remote from ours than the 'language' of our cousin, the dolphin, is remote from our language. Perhaps reluctantly, we must face the fact that on a cosmic scale the 'hard' physical sciences have something of the same cultural relativity that one encounters with the materials of the 'softer' social sciences on a terrestrial basis.

We must accept, with respect to natural science, a rough functional equivalent of the 'historicist' position, urged by Wilhelm Dilthey and others, with respect to history. Science is a cultural artifact whose status is akin to that of the other such artifacts in point of change and transience. We can attain no Hegelian 'position of reason itself' outside of, and contradistinguished from, the fallible and imperfect positions that constitute the changing sequence of historical stages. Such progress as the development of science affords – and substantial it is – is not a matter of discernible evolution toward some history-external, transcendental, position, but a matter of exchanging for one imperfect and ephemeral position yet another position which, while indeed having some significant advantage over its predecessor in point of its substantiating warrant, is for all that itself no more ultimate or definitive. And what holds here with respect to *historical* changes may be presumed to hold with respect to the *situational* changes of mutually remote civilizations as well.

Each inquiring civilization must be expected to produce its own, perhaps ever-changing cognitive product – more or less adequate in its own way but with little if any mutual overlap in conceptual content. Human organisms are essentially similar, but there is not much similarity between the medicine of the ancient Hindus and that of the ancient Greeks. There is every reason to think that the natural science of different astronomically remote civilizations should be highly diversified. Even as different creatures can have a vast variety of life-styles for adjustment within one selfsame physical environment like this earth, so they can have a vast variety of thought-styles for cognitive adjustment within one selfsame world.

With respect to his hypothetical Planetarians, the ingenious Christiaan Huygens wrote, three centuries ago:

> 'Well, but allowing these Planetarians some sort of reason, must it needs be the same with ours? Why truly I think 'tis, and must be so; whether we consider it as applied to Justice and Morality, or exercised in the Principles and Foundations of Science . . . . For the aim and design of the Creator is every where the preservation and safety of his Creatures. Now when such a reason as we are masters of, is necessary for the preservation of Life, and promoting of Society (a thing that they be not without, as we shall show) would it not be strange that the Planetarians should have such a perverse sort of Reason given them, as would necessarily destroy and confound what it was design'd to maintain and defend? But allowing Morality and Passions with those Gentlemen to be somewhat different from ours . . . yet still there would be no doubt, but that in the search after Truth, in judging of the Consequences of things, in reasoning, particularly in that fort which belongs to Magnitude or Quantity, about which their Geometry (if they have such a thing) is employ'd, there would be no doubt I say, but that their Reason here must be exactly the same, and go the same way to work with ours, and that what's

true in one part will hold true over the whole Universe; so that all
the difference must lie in the degrees of Knowledge, which will be
proportional to the Genius and Capacity of the Inhabitants?'
(Huygens 1698, pp. 41–3.)

With the exception of a shift from a theological to a natural-selectionist
rationale, this argumentation could be advanced today. It seems reasonable to argue: 'Common problems constrain common solutions. Intelligent alien civilizations have in common with us the problem of cognitive
accommodation to a shared world. Natural science as we know it is *our*
solution of this problem. *Ergo* it is likely to be *theirs* as well.' But this
tempting argument founders on its second premiss. Their problem is not
common with ours because their situation must be presumed to be
substantially different, since they live in a significantly different environment and come equipped with significantly different resources. To presuppose a common problem is in fact to beg the question.

In philosophical perspective, the 'one world, one science' doctrine
reflects a Hegelian monism. Its basis is the idea that there is one single
'position of reason alone,' and that in the course of cognitive evolution,
nature, metaphorically speaking, works towards a self-knowledge of its
fundamental principles, evolving beings capable of cognitively inter-
nalizing its *modus operandi* at the level of the rational principles. The
fundamental drive is that of cognitive unification in a single, uniquely
determined rational system. The pluralism of a William James stands in
opposition to all this. Its basis is the idea that in the course of cognitive
evolution, nature, metaphorically speaking, works towards being known
in all its various aspects, evolving beings capable of comprehending it in
many different and diversified modes. Here we are dealing with a
teleology of diversity – a nature striving to be known in a variety of forms
and aspects. The positions being weighed in our discussion reflect the
rivalry of these two philosophical metaphors.

And here the advantage clearly lies in the pluralistic side. The one-
world, one-science argument would only go through if it could be
maintained that, while the process or course of scientific development is
something variable and contingent, the ultimate product that will issue
from these diversified strivings is fixed in preordained uniformity. It
would have to be shown that here all different routes lead inexorably and
inevitably to the same destination. But this thesis clearly represents a
problematic and implausible bit of metaphysics.

Our science reflects not only our interests but also our capacities. It
addresses a range of issues that are correlative without specific modes of
physical interaction with nature, the specific ways in which we monitor its

processes. It is selective – the science of a being that gets its most crucial information through sight, monitoring developments along the spectrum of electromagnetic radiation, rather than, say, monitoring variations of pressure or temperature. It is certainly not a phenomenalistic science geared to the feel of things or the taste of things. The science we have developed depends on our capacities and needs, our evolutionary heritage as beings inserted into the orbit or natural phenomena in a certain particular way.

The fact that all intelligent beings inhabit the same world does not countervail against the no less momentous fact that we inhabit very different ecological niches within it, which bring about different sorts of *modus operandi* – physically and cognitively as well, since we cannot separate *theoria* from *praxis*. The chain that links cognition to power, power to a power base, and a power base to a mode of enmeshment in nature, is wrought of unbreakable links.

All such factors as capacities, requirements, interests and course of development affect the shape and substance of the science and technology of any particular place and time. Unless we narrow our intellectual horizons in a parochially anthropomorphic way, we must be prepared to recognize the great likelihood that the 'science' and 'technology' of another civilization will be something *very* different from science and technology as we know it. We are led to view our human sort of natural science as *sui generis*, adjusted to and coordinate with a being of our physical constitution, inserted into the orbit of the world's processes and history in our sort of way. It seems that in science as in other areas of human endeavor we are prisoners of the thought world that our biological and social and intellectual heritage affords us.

Philosophers have often made statements to the effect that people whose experience of the world is substantially different from our own are bound to conceive of it in very different terms. Sociologists, anthropologists, and linguists say much the same sort of things, and philosophers of science have recently also come to talk in this way. According to Thomas Kuhn, for example, scientists who work within different scientific traditions – and thus operate with different descriptive and explanatory 'paradigms' – actually 'live in different worlds' (Kuhn 1962, chap. X).

Supporting considerations for this position have been advanced from very different points of view. One example is a *Gedankenexperiment* suggested by Georg Simmel in the last century – that of envisaging an entirely different sort of cognitive being, intelligent and actively inquiring creatures (animals, say, or beings from outer space) whose experiential modes are quite different from our own (Simmel 1895, pp. 40–1.) Their senses respond rather differently to physical parameters – relatively insensitive, perhaps, to heat and light, but substantially sensitized to various electromagnetic phenomena. Such intelligent creatures, Simmel held, could plausibly be supposed to operate within a largely different framework of empirical concepts and categories – the events and objects of the world of their experience might be very different from those of our own – their phenomenological predicates, for example, might have altogether variant descriptive domains. In a similar view, William James wrote:

> 'Were we lobsters, or bees, it might be that our organization would have led to our using quite different modes from these [actual ones] of apprehending our experiences. It *might* be too (we cannot dogmatically deny this) that such categories un-imaginable by us to-day, would have proved on the whole as serviceable for handling our experiences mentally as those we actually use.'
> (James 1907, p. 171.)

Different cultures and different intellectual traditions, to say nothing of different sorts of creatures, are bound to describe and explain their experience – their world as they conceive of it – in terms of concepts and categories of understanding substantially different from ours. They would proceed with radical divergence with respect to what the Germans call their *Denkmittel* – the conceptual instruments they employ in thought about the facts (or purported facts) of the world. They could, accordingly, be said to operate with different conceptual schemes, with different conceptual tools used to 'make sense' of experience – to characterize, describe, and explain the items that figure in the world as they view it. The taxonomic and explanatory mechanisms by means of which their cognitive business is transacted might differ so radically from ours that intellectual contact with them would be difficult or impossible.

The one world, one science argument shatters against the fact that it is different *thought-worlds* that are at issue in the elaboration of a 'science.' There is no quarrel here with 'the principle of the uniformity of nature.' But this principle merely tells us that, when exactly the same question is put to nature, exactly the same answer will be forthcoming. However, the development of a science hinges crucially on this matter of questions – to the sorts of issues that are addressed and the sequential order in which they are posed. And here the prospect of variation arises: we must expect alien beings to question nature in ways very different from our own. On the basis of an *interactionist* model, there is no reason to think that the sciences of different civilizations will exhibit anything more than the roughest sorts of family resemblance to each other.

A quantitative perspective

Let us attempt to give quantitative substance to the preceding qualitative deliberations.

First, we have the problem of estimating H, the number of habitable planets in the universe. This estimate can be formed by means of the following quantities, themselves represented merely as order-of-magnitude specifications:

n_1 = number of galaxies in the observable universe (10^{11})

n_2 = average number of star systems per galaxy (10^{11})

x_1 = fraction of star systems having suitably large and stable planets (1/10)

n_3 = average number of such planets in the temperate zone of a solar system, where it is neither too hot nor too cold for life (1).

x_2 = fraction of temperate planets equipped with a surface chemistry capable of supporting life (1/10)

These figures – borrowed in the main from Dole (1970) and Sagan (1980) – ought to be viewed with scepticism. They, and those that are to follow, must be taken realistically. They are not graven in stone for all the ages, but represent reasonable 'best estimates' in the present state of the art. The important point, as will emerge below, is that the overall tendency of our discussion is not acutely sensitive to precision in this respect. One should look on the calculations that are to follow as suggestive rather than in any sense conclusive. Their function is to motivate a tendency of thought, and not to establish a definitive conclusion.

Given the preceding estimates, the sought-for number of habitable planets will be the product of these quantities:

$$H = 10^{20}$$

This, of course, is a prodigiously large number, providing for some thousand million habitable planets per galaxy (cf. Dole 1970, p. 103). But this is only the start of the story.

A planet capable of supporting life might well have no life to support, let alone *intelligent* life. The point is that the physics, chemistry, and biology must all work out just right. The physical, chemical, and biological environments must all be duly auspicious, and exactly the right course of triggering processes must unfold for the evolution of intelligence to run a successful course.

Our next task is thus to estimate I, the number of planets on which intelligent life evolves. Let us proceed here via the following (admittedly rough and ready) quantities:

r_1 = fraction of habitable planets on which life – that is, *some sort* of self-reproducing biological system – actually arises. (1/100)

r_2 = fraction of these on which conditions are sufficiently favorable that highly complex life-forms evolve, possessed of something akin to a central nervous system and thus capable of complex (though yet instinctively programmed) behavior forms. (1/100)

r_3 = fraction of these on which intelligent and sociable beings evolve – beings who can acquire, process, and exchange factual information with relative sophistication – who can observe, remember, reason, and communicate. (1/100)

As these fractions indicate, the evolutionary process that begins with the inauguration of life and moves on to the development of intelligence is not an inexorable sequence but one which could, given suitably inauspicious conditions, abort in a stabilization that freezes the whole course of development at some plateau along the way. Note that r_3 in particular involves problems. Conscious and indeed even intelligent creatures are readily conceivable who yet lack that orientation towards their environment needed to acquire the factual information necessary to science. Where such conceptions as space, time, process, unit, function, and order are missing, it is difficult to see how anything deserving of the name 'science' could exist. Then too, an intelligence unswervingly preoccupied with the aesthetic appreciation of particular phenomena rather than their generally lawful structure is going to miss out on the scientific dimension.

When we put the fractions of the preceding series to work, we arrive at

$$I = 10^{14}$$

This unquestionably indicates an impressively large number of intelligence-bearing planets. It would, in fact, yield a quota of some thousand per galaxy (a figure which if correct, would cast a shadow over the prospect of our ever establishing contact with extraterrestrial intelligence, since it would indicate its nearest locale to be some 1000 light-years away). Note that this is a rather middle-of-the-road figure. Some see our galaxy teeming with a million intelligence-bearing planets. Others (e.g.

Trefil in Rood & Trefil 1981) think that ours is probably the only planet in the galaxy inhabited by intelligent creatures.

This number, of course, rests on a shaky foundation. One is staggering about in the dark, chasing data one cannot see, playing a quasi-scientific blind-man's buff. (The estimates in the published literature of the number of intelligence-bearing planets in our galaxy range from one upwards into many millions.) As regards our own figure, one can say that it would certainly be possible to take a more rosy view of the matter. One could suppose that nature has a penchant for life – that a kind of Bergsonian *élan vital* is operative, so that life springs forth wherever it can possibly get a foothold. Something of this attitude certainly underlies J. P. T. Pearman's contention (Cameron 1963, p. 290) that the probability is 1 that life will develop on a planet with a suitable environment – a stance in which MacGowan & Ordway 1966 (p. 365), Dole 1970 (pp. 99–100), and Ball 1973 (p. 347) concur. (Sagan 1980, p. 300, is slightly more conservative in fixing this quantity at 1/3.) One theorist cuts the Gordian Knot with a curious bit of reasoning:

> 'Biological evolution proceeds by the purely random process of mutation . . . . Since the process is a random one, the laws of probability suggest that the time-scale of evolution on earth should resemble the average time-scale for the development of higher forms of life anywhere.'
> (Huang 1960, p. 55)

But this ignores the crucially differentiating role of initial conditions in determining the outcome of random processes. The terrain through which a random walk proceeds is going to make a lot of difference to its destination. The transition from habitability to habitation – from the possibility of life to its actuality – is surely not all that simple. Sir Arthur Eddington did well to remind us in this context of the prodigality of nature when he asked how many acorns are scattered for any one that grows into an oak (Eddington 1928, p. 177).

One could perhaps suppose that nature incorporated a predisposition for intelligence – that there is an impetus, reminiscent of that of Teilhard de Chardin, towards *nous*, so that intelligence develops wherever there is life. ('[T]he evolution of a human (or a space communication) level of intelligence is inevitable on a planet when many varied species flourish over a sufficiently long period of time.' MacGowan & Ordway 1966, p. 365.) Indeed the suggestion is sometimes made in this vein that 'the adaptive value of intelligence . . . is so great . . . that if it is genetically feasible, natural selection seems likely to bring it forth' (Shklovskii &

Sagan 1966, p. 411). But this argument from utility to evolutionary probability clearly has its limitations. ('[T]here are no organisms on Earth which have developed tractor treads for locomotion, despite the usefulness of tractor treads in some environments' *Ibid.*, p. 359.) Moreover, this suggestion seems implausibly anthropocentric. To all appearances the termite has a securer foothold on the evolutionary ladder than man; and the coelacanth (now some 350 million years old – over 125 million years older than the dinosaurs) can afford to smile when the survival advantages of intelligence are touted by a Johnny-come-lately creature whose self-posed threats to long-term survival are nowadays a cause of general concern. As J. P. T. Pearman has rightly noted, 'the successful persistence of a multitude of simpler organisms from ancient times argues that intelligence may confer no unique benefits for survival in an environment similar to that of earth' (Cameron 1963, p. 290). For creatures with the luck to possess an instinct for constructing a viable environment – or the even better luck to find one ready-made in every relevant respect – intelligence loses its biological and evolutionary value. It will prove survival-conducive mainly for a being of a particularly restless disposition – a creature like man, who refuses to settle down in a secured ecological niche, but shifts restlessly from environment to environment needing continually to readjust to self-imposed changes. The value of intelligence, one might say, is not absolute but remedial – to offset the problems of a particular sort of life-style.

The number of possible pathways that an evolutionary course might take is enormous, and the probability of an organism evolving with a central nervous system like man's is extremely small (cf. Beadle 1960.) In sum, we do well to think of the emergence of intelligence as a long series of fortuitous twists and turnings rather than an inexorable push towards a foreordained result. It is unrealistically egocentric to regard intelligence as an inevitable fruit of organic evolution: the evolutionary course from life to intelligence follows the contingent twists and turns of an incredibly complex route and it would be glib in the extreme to assume that, once life arises, its subsequent development would proceed in much the same way as here on earth. Simpson (1964) provides a useful perspective here.

The indicated figures accordingly seem plausibly middle-roadish between pessimism and an intelligence-favoring optimism that does not seem warranted at this particular stage of the scientific game. Even so, it is clear that the proposed specification of *I* represents a strikingly substantial magnitude – one which contemplates many thousands of millions of planets equipped with intelligent creatures scattered throughout the universe.

Intelligence, however, is not yet the end of the line. (After all, dolphins and apes are presumably intelligent, but they do not have a 'science.') Many further steps are needed to estimate S, the number of planets throughout the universe in which scientific civilizations arise. The developmental path from intelligence to science is a road strewn with substantial obstacles. Here matters must be propitious not just as regards the physics, chemistry, biochemistry, evolutionary biology and cognitive psychology of the situation: The social-science requisites for the evolution of science as a cultural artifact must also be met. Economic conditions, social organization, and cultural orientation must all be properly adjusted before the move from intelligence to science can be accomplished. For scientific inquiry to evolve and flourish there must, in the first place, be cultural institutions whose development requires economic conditions and a favorable social organization. And terrestrial experience suggests that such conditions for the social evolution of a developed culture are by no means always present where there is intelligence. At this point we do well to recall that of the thousands of human civilizations evolved here on earth only one, the Mediterranean/European, managed to develop natural science. The successful transit from intelligence to science is certainly not a sure thing.

Moreover, the development of science is a social process closely tied to a suitably evolving technology that is itself symbiotically bound up with the course of scientific development. And it must be recognized that even if a civilization has technology it need not have science. Its technology might be altogether disconnected from science, having evolved over aeons of time by an immense course of trial and error, without building on a basis of theoretical understanding. A lot of know-how can be built up without much know-why. Again, our aliens' technology might well be social and organizational rather than instrumental. Like colonies of highly versatile ants, they could organize so as to use their 'Natural' instrumentation for a varied, complex, and sophisticated plurality of tasks of physical manipulation without much hardware entering in.

Let us once more look at the matter quantitatively:

p_1 = probability that intelligent beings will (unlike dolphins) also possess developed manipulative abilities and will (unlike the higher apes) combine intelligence with manipulative ability so as to develop a technology that can be passed on as a social heritage across the generations. (0.01)

p_2 = probability that technologically competent intelligent beings will group themselves in organized societies of substantial complexity – a transition that stone-age man, for example, never managed to make. (0.1)

p_3 = probability that an organized society will not only acquire the means for transmitting across successive generations the political and pragmatic 'know how' indispensable to an 'organized society' as such, but will also (unlike the ancient Egyptians) develop institutions of learning and culture for accumulating, refining, systematizing, and perpetuating factual information. (0.1)

p_4 = probability that society with cultural institutions will develop an unstable (i.e. continually developing and dynamic) technology – in the way the ancient Greeks and the old Chinese mandarins, for example, never did – so as to create a technologically progressive civilization. (0.01)

p_5 = probability that a technologically progressive civilization will develop and maintain an articulated 'science' and concern itself with the theoretical study of nature at a level of high generality and precision. (0.1)

However firm the physical quantities with which we began, we are by now skating on very thin ice indeed. These issues of sociology and cognitive psychology can only be quantified in the most tentative and cautious way. But the one thing that is clear is that a good many conditions of this sort have to be met and that each reaches a likelihood of relatively modest proportions.

The issue of technology reflected in p_1 and p_4 is particularly critical here. The urge to an ever more aggressive technological extension of self is certainly not felt by every intelligent life form. It is a part of Western man's peculiar life-style impatiently to cultivate the active modification of nature in the pursuit of human convenience so as to create an artificial environment of ultra-low entropy. Even in human terms this is not a solution uniquely constrained to the problem of evolutionary adaptation. Many human societies seem to have remained perfectly content with the *status quo* for countless generations and very sophisticated cultural projects – literary criticism, for example – have developed in directions very different from the scientific. Anyone who thinks that $p_3 - p_4$ yields too low a probability for the transition from culture to science would do well to reflect on our experience here on earth. A culture can easily settle comfortably into frozen traditionary patterns with respect to technology. (If their attention span is long enough, our aliens might cultivate scholastic theology *ad infinitum*.) Moreover, unless their oral lore is something very different from ours, it is hard to see how an alien civilization could develop science without writing – a skill which even many human communities did not manage to develop. The salient point is that for 'science' to emerge on a distant planet it is not enough for there to be life and intelligence; there must also be culture and progressive technology and explanatory interest and theorizing competency.

The product of the preceding sequence of probability estimates is 10^{-7}. Multiplying this by I, we would obtain the following expected-value estimate of the number of 'science'-possessing planetary civilizations.

$$S = 10^7 = 10,000,000$$

This, of course, is a large number, albeit one that is rather modest on a cosmic scale, implyiing a chance of only some 0.01 % that a given galaxy actually provides the home for a 'science.' (Note too that we ignore the temporal dimension – the scientific civilizations at issue may have been destroyed long ago, or perhaps simply have lost interest in science.) A more conservative appraisal of the sociological parameters has thus led us to a figure that is more modest by many orders of magnitude than the estimate by Shklovskii & Sagan (1966, p. 418) that some $10^{5\pm1}$ scientifically sophisticated civilizations exist in our galaxy alone. Nevertheless, even our modest 10 million is still a very substantial number.

Comparability and judgments of relative advancement or backwardness

Let us now come to grips with the crux of our present concerns: the issue of scientific advancement. Earlier we defined 'science' in terms of a rather generic *functional* equivalency. The question, however, that concerns us here is not whether a remote civilization has a 'science' of some sort, but whether it is *scientifically more advanced* than ourselves. But if *advancement* is to be at issue, and the question is to be one of relative backwardness or sophistication, then an actual substantive comparison must be provided. We are now dealing not with the essentially quantitative aspects of intelligence as such – with data processing in volumetric terms – but with the quality of the orientation of intelligence to substantive issues. If another science is to represent an advance over ours, we must clearly construe it as *our sort* of science in much more particularized and substantive terms. And given the *immense* diversity to be expected among the various modes of 'science' and 'technology,' the number of extraterrestrial civilizations possessing a science and technology that is duly consonant and contiguous with ours – and, in particular, heavily geared towards the mathematical laws of the electromagnetic spectrum – must be judged to be very small indeed.

We have come to recognize that sciences can vary (1) in their formal mechanisms of *formulations* (their 'mathematics'); (2) in their *conceptualization*, i.e. in the kinds of explanatory and descriptive concepts they bring to bear; and (3) in their *orientation* towards the manifold pressures of nature, reflecting the varying 'interest'-directions of their developers. While 'science' as such is clearly not anthropocentric, science *as we have it* – the only 'science' that we ourselves know – is a specifically human artifact that must be expected to reflect in significant degree the particular characteristics of its makers. Consequently, the prospect that an alien 'science'-possessing civilization has a *science* that we would acknowledge (if sufficiently informed) as representing the same general line of inquiry as that in which we ourselves are engaged seems extremely implausible. The possibility that *their* science-and-technology is 'sufficiently similar' in orientation and character to be substantively proximate to *ours* must be viewed as extremely remote. We clearly cannot estimate this as representing something other than a very long shot indeed – certainly no better than one in many thousands.

Just such comparability with 'our sort of science' is, however, the indispensable precondition for judgments of relative advancement or backwardness *vis-à-vis* ourselves. The idea of their being scientifically 'more advanced' is predicated on the uniformity of the enterprises – doing better and more effectively the kinds of things that *we* want science and technology to do. Any talk of advancement and progress is predicated on the sameness of the direction of movement: only if others are traveling along the same route as ourselves can they be said to be ahead of or behind us. The issue of relative advancement is linked inseparably to the idea of doing *the same sort of thing* better or more fully. And this falls apart when 'this sort of thing' is not substantially the same. One can say that a child's expository writing is more primitive than an adult's, or that the novice's performance at arithmetic or piano-playing is less developed than that of the expert. But we can scarcely say that Chinese cookery is more or less advanced than Roman, or Greek pottery than Renaissance glass-blowing. The salient point for present purposes is simply that where the enterprises are sufficiently diverse, the ideas of advancement and progress are inapplicable for lack of a *sine qua non* condition. (Of course, even 'doing the same sort of thing' at a very remote level of generality may not suffice for the comparability judgments of doing it better or worse that are needed to underwrite judgments of relative progressiveness or backwardness. The condition at issue is merely necessary and not sufficient. Butterflies and sparrows both fly, but, given the differences in their *modus operandi*, it makes little sense to say that one does so better than the other.)

Our reasoning here has occasionally used the analogy of spatial and temporal remoteness. But in this context of advancement or progressiveness there now arises a crucial disanalogy between *alien* science and *future* science. Future science is inevitably comparable with ours: it is the successor-state of a successor-state of a successor-state, and advantageousness is transmitted along this chain of succession. If a state of science were not significantly more advanced than one already in place, it would (presumably) not establish itself as a successor. The very conditions of historic change in our terrestrial science assure its progressiveness in point of prediction and control, and thus presumably in point of explanatory adequacy as well. No such in-principle assurance of relative comparability is available in the case of an alien 'science.'

The matter of entitlement to claim scientific superiority is not as simple as may seem at first sight. To begin with, scientific superiority does not automatically emerge from their capacity to make many splendidly successful predictions. For this could be the result of precognition or empathetic attunement to nature or such-like. Again, what is wanted is not just a matter of *correct*, but of cognitively underwritten, and thus *science-guided* predictions. And that is just exactly what is to be proved.

It clearly is not enough for establishing their being scientifically more advanced than ourselves that the aliens should perform 'technological wonders' – that they should be able to do all sorts of things we would like to do but cannot. After all, bees can do that. The technology at issue must clearly be the product of intelligent contrivance rather than evolutionary trial and error. What is needed for advancement is that their performatory wonders issue from superior theoretical knowledge – i.e. from superior science. And then we are back in the circle.

Nor would the matter be settled by the consideration that an extra-terrestrial species might be more 'intelligent' than ourselves in having a greater capacity for the timely and comprehensive monitoring and processing of information. After all, whales or porpoises, with their larger brains, may (for all we know) have to manipulate relatively larger quantities of sheer data than we do to maintain effective adaptation within a highly changeable environment. What clearly counts for scientific knowledge is not the *quantity* of intelligence in sheer volumetric terms but its *quality* in substantive, issue-oriented terms. Information handling does not assure scientific development. Libraries of information (or misinformation) can be generated about trivia – or dedicated to matters very different from science as we know it.

It is perhaps too tempting for humans to reckon cognitive superiority by the law of the jungle – judging as superior those who do or would come out on top in outright conflict. But surely the Mongols were not possessors of a civilization superior to that of Near-Eastern cultures they overran. Again, we earthlings might easily be eliminated by not-very-knowledgeable creatures able to produce at will – perhaps by using natural secretions – a biological or chemical agent capable of killing us off.

The key point then, is that if they are to effect an *advance* on our science they must both (1) do roughly our sort of thing in roughly our sort of way, and (2) do it significantly better. In speaking of the 'science' of another civilization as 'more advanced' than our own we contemplate the prospect that they have developed *science* (*our* sort of science – 'science' as we know it) further than we have ourselves. And this is implausible. Even assuming that 'they' develop a 'science' at all – that is, a *functional equivalent* of our science – it seems unduly parochial to suppose that they are at work constructing *our* sort of science in substantive, content-oriented terms. Diverse life-modes have diverse interests; diverse interests engender diverse technologies; diverse technologies make for diverse modes of science.

If a civilization of intelligent aliens develops a science at all, it seems plausible to expect that they will develop it in another direction altogether and produce something that we, if we could come to understand it at all, would regard as simply detached in content-orientation – though perhaps not in intent – from the scientific as we ourselves cultivate it. (Think of the attitude of orthodox sciences to 'exotic' phenomena like hypnotism, let alone to parapsychology.)

It is naive to think of the relationship of different modes of 'science/technology' in the sort of terms formulated from such comparisons as higher/lower, inferior/superior, crude/sophisticated, primitive/advanced. This hierarchical thinking is already in difficulty when applied to the cultural products of terrestrial civilizations. It is in yet deeper difficulty when applied to the cognitive product of astronomically remote civilizations where an infinitely greater variability of aspects can be envisaged.

There just is no single-track itinerary of scientific/technological development that different civilizations travel in common, with mere differences in speed or in staying power (notwithstanding the penchant of astrophysicists for the neat plotting of numerical 'degrees of development' against time in the evolution of planetary civilizations – cf. Ball 1980, p. 658). In cognition and even in 'scientific' evolution we are not dealing with a single-track railway, but with a complex network leading to many mutually remote destinations. Even as cosmic evolution involves a red shift that carries different star systems even farther from each other, so cognitive evolution may well involve a red shift that carries different civilizations ever farther from each other into mutually remote thought-worlds.

The literature generated by extraterrestrial intelligence enthusiasts is fraught with the haunting worry: 'Where is everybody?' 'Why haven't we heard from them?' Are they simply too distant – or perhaps too cautious (see Nozick 1972) or too detached (see Ball 1973)? Our present discussion offers yet another line of response: they are simply too busy doing their own thing. Radiocommunication is ours, theirs is something very, *very* different. Alien civilizations inhabit alien thought-worlds, and lack of intellectual communion engenders lack of physical communication.

The prospect that an alien civilization is going about the job of doing *our* science – a 'science' that reflects the sorts of interests and involvement that *we* have in nature – better than we do ourselves must accordingly be adjudged as extremely far-fetched. Specifically two conditions would have to be met for the science of an intelligent civilization to be in a position to count as comparable to ours.

(1) That, *given* that they have a 'science' and a developing 'technology,' they have managed to couple the two and have proceeded to develop (unlike the ancient Greeks, the Chinese and the Byzantines) a *science-guided* technology. (Probability p_6.)

(2) That their science-guided technology is sufficiently oriented towards issues regarding natural processes sufficiently close to those at which our science-guided technology is oriented that a comparison can reasonably be made between them. (Probability p_7.)

To judge by terrestrial experience, it seems rather optimistic to estimate p_6 to be even so large as one in a thousand. And p_7 must also be adjudged as quite small. As we have seen, science-guided technology could be oriented in very different directions. The potential diversity of different modes of 'science' is enormous; there is little choice but to see p_7 as an eventuation whose chances are no better than, say, one in ten thousand (so $p_7 = 0.0001$). If our alien scientists are differently constructed (if they are silicon-based creatures, for example), or if their natural environment is very different, their practical interests and its accordant technology will be oriented in very different directions from ours. Even we ourselves might well have aimed our technological development not at 'mastery over external nature' but at the advancement of 'control over oneself' in the manner of some oriental civilizations. Analogously, their technology might be wholly independent of 'hardware,' oriented not towards physical machinery, but towards the software of mind-state manipulation, telepathy, hypnotism, auto-suggestion, or the like. (Ray Bradbury's Martians use thought-control to destroy an expedition from earth armed with atomic weapons (Bradbury 1954).)We must not keep our imagination on a short leash in this regard. Given the diversity of different modes of 'science' and the enormous spectrum of possible issues and purposes in principle available to extra-terrestrial aliens, the prospect must be recognized that the direction of their science-guided technology might be vastly different from our own.

Accordingly we have it that

$$p_6 \times p_7 = 10^{-7}$$

The product of this quantity with S, the number of civilizations that possess a technological science as we comprehend it, is clearly not going to be very substantial – it is, in fact, going to be strikingly close to 1.

If 'being there' in scientific regards means having *our* sort of scientifically guided technology and our sort of technologically channeled science, then it does not seem all that far-fetched to suppose that *we might be there alone* – even in a universe teeming with other intelligent civilizations. The prospect that somebody else could do 'our sort of thing' in the scientific sphere better than we can do it ourselves seems very remote.

First principles

The overall structure of our analysis thus emerges in the picture of Table 1. Its figures interestingly embody the familiar situation that as one moves along a nested hierarchy of increasing complexity one encounters a greater scope for diversity – that the further layers of system complexity provide for an ever widening spectrum of possible states and

Table 1. *Conditions for the development of science*

Planets of sufficient size for potential habitation	10^{22}
Fraction thereof with the right:	
astrophysics for a temperate location	10^{-1}
chemistry for life-support	10^{-1}
biochemistry for the actual emergence of life	10^{-2}
biology and psychology for the evolution of intelligence	10^{-4}
sociology for developing a culture with a 'technology' and a 'science'	10^{-7}
epistemology for developing science as we know it	10^{-7}

conditions. (The more fundamental the system, the narrower its correlative range of alternatives; the more complex, the wider.) If each unit ('letter,' 'cell,' 'atom') can be configurated in 10 ways, then each ordered group of 10 such units ('word,' 'organ,' 'molecule,') can be configurated in 10^{10}, and each complex of 10 such groups ('sentences,' 'organisms,' 'objects') in $(10^{10})^{10} = 10^{100}$ ways. Thus even if only a small fraction of what is realizable in theory is realizable in nature, any increase in organizational complexity will nevertheless be accompanied by an enormous amplification of possibilities.

To be sure, the specific particulars of the various computations that have formed the quantitative threat of the discussion cannot be given much credence. But their general tendency nevertheless conveys an important lesson. For people frequently seem inclined to reason as follows:

> There are, after all, an immense number of planetary objects running about in the heavens. And proper humility requires us to recognize that there is nothing all that special about the earth. If it can evolve life and intelligence and civilization and science, then so can other planets. And given that there are so many other runners in the race we must assume that – even though we cannot see them in the cosmic darkness – some of them have got ahead of us in the race.

As one recent writer formulates this familiar argument, 'Since man's existence on the earth occupies but an instant in cosmic time, surely intelligent life has progressed far beyond our level on some of these 100 000 000 (habitable) planets (in our galaxy)' (M. Calvin in Cameron 1963, p. 75). But so plausible-sounding an argument overlooks the numerical complexities. Even though there are an immense number of solar systems, and thus a staggering number of planets (some 10^{22} on our estimate), nevertheless, a substantial number of conditions must be met for 'science' (as we understand it) to arise. The astrophysical, physical, chemical, biological, psychological, sociological, and epistemological parameters must all be in proper adjustment. There must be habitability, life, intelligence, culture, technology, a 'science' coupled to technology, an appropriate subject-matter orientation of this intellectual product, etc. A great many turnings must go aright *en route* to science of a quality comparable to ours. Each step along the way is one of finite (and often smallish) probability. And, to reach the final destination, all these probabilities must be multiplied together, yielding a quantity that might be very small indeed. Even if there were only 12 turning points along this developmental route, each involving a chance of successful eventuation that is on average no worse than a one-in-a-hundred, the chance of an overall success would be immensely small, corresponding to an aggregate success-probability of merely 10^{-24}.

It is tempting to say 'The universe is a big place; surely we must expect that what happens in one locality will be repeated someplace else.' But this overlooks the issue of probability. Admittedly cosmic locales are very numerous. But probabilities can get to be very small: no matter how massive N, there is that diminutive $1/N$ that can countervail against it.

In the course of evolutionary development things can eventuate very differently at many junctures. George G. Simpson has rightly stressed the many chancy twists and turns that lie along the route, insisting that:

> 'the fossil record shows very clearly that there is no central line leading steadily, in a goal-directed way, from a protozoan to man. Instead there has been continual and extremely intricate branching, and whatever course we follow through the branches there are repeated changes both in the rate and in the direction of evolution. Man is the end of one ultimate twig . . . . Even slight changes in earlier parts of the history would have profound cumulative effects on all descendant organisms through the succeeding millions of generations . . . . The existing species would surely have been different if the start had been different, and if any stage of the histories of organisms and their environments had been different. Thus the existence of our present species depends on a very precise sequence of causative events through some two billion years or more. Man cannot be an exception to this rule. If the causal chain had been different, *Homo sapiens* would not exist.'
> (Simpson 1964, p. 773.)

The workings of evolution – be it of life, intelligence, culture, technology, or science – are always the product of a great number of individually unlikely events. The unfolding of developments involves putting to nature a series of questions whose successive resolution produces a process reminiscent of the game '*Twenty Questions*,' sweeping over a possibility-spectrum of awesomely large proportions. The result eventually reached lies along a route that traces out one particular contingent path within a space of alternatives that provides for an ever-divergent fanning out of alternatives as each step opens up yet further possibilities. An evolutionary process is a very 'iffy' proposition – a complex labyrinth where a great many twists and turns in the road must be taken aright for matters to end up as they do.

Of course, it all looks easy with the wisdom of hindsight. If things had not turned out appropriately at every stage, we would not be here to tell the tale. The many contingencies on the long route of cosmic, galactic, solar-systemic, biochemical, biological, social, cultural, and cognitive evolution have all turned out aright – the innumerable obstacles have all been surmounted. In retrospect it all looks easy and inevitable. The innumerable possibilities of variation along the way are easily kept out of sight and out of mind. The Whig interpretation of history beckons comfortably. It is so easy, so tempting, to say that a planet on which there is life will of course evolve a species with the technical capacity for interstellar communication. (Cf. Cameron in Cameron 1963, p. 312, who fixes this conditional probability as 1.) It is tempting, but it is also nonsense. For the fact remains that, at many critical turnings on evolution's road, we face the circumstance that had things gone only a little differently, we would not be here. (Cf. Rood & Trefil 1981, who develop this line of thought in cogent and plausible detail.)

The ancient Greek atomists' theory of possibility affords an interesting object-lesson in this connection. Adopting a Euclideanly infinitistic view of space, they held to a theory of innumerable worlds:

> 'There are innumerable worlds, which differ in size. In some worlds there is no sun and moon, in others they are larger than in our world, and in others more numerous. The intervals between the worlds are unequal; in some parts there are more worlds, in others fewer; some are increasing, some at their height, some decreasing; in some parts they are arising, in others failing. They are destroyed by collision one with another. There are some worlds devoid of living creatures or plants or any moisture.' (Diels-Kranz 68 A 40 (for Leucippus and Democritus); Kirk & Raven 1957, p. 411.)

On this basis the atomists taught that every (suitably general) possibility is realized in fact someplace or other. Confronting the question of 'Why do dogs not have horns: just why is the theoretical possibility that dogs be horned not actually realized?' the atomists replied that it indeed is realized but just elsewhere – *in another region of space.* Somewhere within infinite space there is another world just like ours in every respect save one, that its dogs have horns. For the circumstance that dogs lack horns is simply a parochial idiosyncrasy of the particular local world in which we interlocutors happen to find ourselves. Reality accommodates all possibilities of worlds alternative to this one through spatial distribution: as the atomists saw it, *all* alternative possibilities are in fact actualized in the various subworlds embraced within one spatially infinite superworld.

This theory of virtually open-ended possibilities was shut off by the closed cosmos of the Aristotelian world picture, which dominated European cosmological thought for almost two millennia. The break-up of the Aristotelian model in the Renaissance and its replacement by the 'Newtonian' model is one of the great turning points of the intellectual tradition of the West – elegantly portrayed in Alexandre Koyré's book of the splendid title *From the Closed World to the Infinite Universe* (New York: Harper, 1957). One may recall Giordano Bruno's near-demonic delight with the break-up of the closed Aristotelian world into one opening into an infinite universe spread throughout endless spaces. Others were not delighted but appalled – John Donne spoke of 'all cohearence lost,' and Pascal was frightened by 'the eternal silence of infinite spaces' of which he spoke so movingly in the *Pensées* (§§ 205–206). But no one doubted that the onset of the 'Newtonian' world-picture represented a cataclysmic event in the development of Western thought. Strangely enough, the refinitization of the universe effected by Einstein's general relativity produced scarcely a ripple in philosophical or theological circles, despite the immense stir caused by other aspects of the Einsteinean revolution. (Einsteinean space–time is, after all, even more radically finitistic than the Aristotelian world-picture, which left open at any rate the prospect of an infinite future with respect to time.)

To be sure, it might well seem that the finitude in question is not terribly significant because the distances and times involved in modern cosmology are so enormous. But this view is rather naive. The difference between the finite and the infinite is as big as differences can get to be. And it represents a difference that is – in this present context – of the most far-reaching significance. For this means that we have no alternative to supposing that a highly improbable set of eventuations is not going to be realized in very many places, and that something sufficiently improbable may well not be realized at all. The decisive *philosophical* importance of cosmic finitude lies in the fact that in a finite universe only a finite range of alternatives can be realized. A finite universe must 'make up its mind' about its contents in a far more radical sense than an infinite one. And this is particularly manifest in the context of low-probability possibilities. In a finite world – unlike an infinite one – we cannot avoid supposing that a prospect that is sufficiently unlikely is simply not going to be realized at all, that in piling improbability on improbability we eventually outrun the reach of the actual. It is, accordingly, quite conceivable that our science represents a solution of the problem of cognitive accommodation that is terrestrially locale-specific.

Here lies a deep question. Is the mission of intelligence in the cosmos uniform or diversified? Two fundamentally opposed philosophical positions are possible with respect to cognitive evolution in its cosmic perspective. The one is a uniformitarian *monism* which sees the universal mission of intelligence in terms of a certain shared destination, a common cosmic 'position of reason as such.' The other is a particularistic *pluralism* which allows each solar civilization to forge its own characteristic cognitive destiny, and sees the mission of intelligence as such in terms of spanning a wide spectrum of alternatives and realizing a vastly diversified variety of possibilities, with each thought-form achieving its own peculiar destiny in separation from all the rest. The conflict between these doctrines must in the final analysis be settled not by armchair speculation from general principles but by rational triangulation from the empirical data. This said, it must, however, be avowed that the whole tendency of these present deliberations is towards the more pluralistic side.

The implausibility of being outdistanced

For many there is, no doubt, a certain charm to the idea of companionship. It would be comforting to think that, however estranged we are in other ways, those alien minds and ourselves share *science* at any rate – that we are fellow travellers on a common journey of inquiry. Our yearning for companionship and contact runs deep. (Pascal was not the only one frightened by the eternal silence of infinite spaces.) It would be pleasant to think ourselves not only colleagues but junior collaborators whom other, wiser minds might be able to help along the way. Even as many in sixteenth-century Europe looked to those strange pure men of the Indies (East or West), who might serve as moral exemplars for sinful European man, so we are tempted to look to alien inquirers who surpass us in scientific wisdom and might assist us in overcoming our cognitive deficiencies. The idea is appealing but it is also, alas, very unrealistic.

In the late 1600s Christiaan Huygens wrote:

> 'For 'tis a very ridiculous opinion that the common people have got among them, that it is impossible a rational Soul should dwell in any other shape than ours . . . . This can proceed from nothing but the Weakness, Ignorance, and Prejudice of Men; as well as the humane Figure being the handsomest and most excellent of all others, when indeed it's nothing but a being accustomed to

that figure that makes me think so, and a conceit . . . that no shape
or colour can be so good as our own.'
(Huygens 1698, pp. 76–7.)

What is here said about people's tendency to place all rational minds into
a physical structure akin to their own familiar one is paralleled by a
tendency to place all rational knowledge into a cognitive structure akin to
their own familiar one.

It is clear that life on other worlds might be very different from the life
we know. It might be based on a multivalent element other than carbon
and be geared to a medium other than water – perhaps even one that is
solid or gaseous rather than liquid. In his splendid book *The Immense
Journey* (New York: Random House, 1937) Loren Eiseley wrote:

> 'Life, even cellular life, may exist out yonder in the dark. But
> high or low in nature, it will not wear the shape of man. That
> shape is the evolutionary product of a strange, long wandering
> through the attics of the forest roof, and so great are the chances
> of failure, that nothing precisely and identically human is likely
> ever to come that way again.'

What is here written regarding the material configuration of the human
shape would seem no less applicable to the cognitive configuration of
human thought – scientific thought nowise excluded. The physics of an
alien civilization need resemble ours no more than does their physical
therapy. We must be wary of *cognitive* anthropomorphism just as much as
we are of structural anthropomorphism – they are certainly not 'writ
larger' in the inevitable scheme of things. Immanuel Kant's insight holds
– there is good reason to think that natural science as we know it is not
something universally valued for all rational intelligences as such, but a
man-made creation correlative with our specifically human intelligence.

With respect to biological evolution it seems perfectly sensible to
reason as follows:

> 'What can we say about the forms of life evolving on these other
> worlds? . . . [I]t is clear that subsequent evolution by natural
> selection would lead to an immense variety of organisms; com-
> pared to them, all organisms on Earth, from molds to men, are
> very close realtions.'
> (Shklovskii & Sagan 1966, p. 350.)

It is plausible that much the same situation should obtain with respect to
cognitive evolution: that the 'sciences' produced by different civilizations
here on earth – the ancient Chinese, Indians, and Greeks for example –
should exhibit an infinitely greater similarity than obtains between our

present-day science and anything devised by astronomically remote civilizations. The idea of a comparison in terms of 'advance' or 'backwardness' would simply by inapplicable.

The upshot of these deliberations is thus relatively straightforward. The prospect that some astronomically remote civilization is 'scientifically more advanced' than ourselves – that somebody else is doing 'our sort of science' *better* than we ourselves – requires in the first instance that they be doing our sort of science at all. And this deeply anthropomorphic supposition is extremely unlikely. To endorse the idea of a scientifically superior extraterrestrial civilization is to step beyond the realm of scientifically plausible fact. The quest for superhuman science should bring home to us the uniqueness of that artifact which is our human science. It should be stressed, however, that this consideration that 'our sort of science' may well be unique is not so much a celebration of our intelligence as a recognition of our peculiarity.

The prospect that an alien civilization might carry *science* – i.e. *our* sort of science – further than ourselves is, to put it mildly, remote. Aliens might well surpass us in many ways – in power, in longevity, in intelligence, in ferocity, etc. But to worry (or hope) that they might surpass us in *science as we understand it* is to orient one's concern in an unprofitable direction.

References

This listing is confined to materials that I have found particularly interesting or useful. It does not aspire to comprehensiveness. A much fuller bibliography is given in McGowan & Ordway 1966.

Allen, T. B. (1965). *The Quest: A Report on Extraterrestrial Life.* Philadelphia: Chilton Books. (An imaginative survey of the issues.)

Anderson, P. (1963). *Is There Life on Other Worlds?* New York and London: Collier-Macmillan. (Chapter 8, 'On the Nature and Origin of Science', affords many perceptive observations.)

Ball, J. A. (1973). The Zoo Hypothesis. *Icarus*, **19**, 347–9. (Aliens are absent because the Intergalactic Council has designated earth a nature reserve.)

— (1980). Extraterrestrial intelligence: Where is everybody? *American Scientist*, **68**, 56–63.

Beadle, G. W. (1960). The place of genetics in modern biology. *Engineering and Science*, **23** (6), 11–17. (An insightful analysis of the evolutionary process.)

Beck, L. W. (1971–2). Extraterrestrial intelligent life. *Proceedings and Addresses of the American Philosophical Association*, **45**, 5–21. (A thoughtful and very learned discussion.)

Berrill, N. J. (1964). *Worlds Without End.* London: Macmillan. (A popular treatment.)

Bracewell, R. N. (1975). *The Galactic Club: Intelligent Life in Outer Space.* San Francisco: W. H. Freeman. (A lively and enthusiastic survey of the issues.)

Bradbury, R. (1954). *The Martian Chronicles*. New York, Bantam Books. (Reprinted 1979.)

Cameron, A. G. W. (ed.) (1963). *Interstellar Communication: A Collection of Reprints and Original Contributions*. New York and Amsterdam: W. A. Benjamin. (A now somewhat dated but still indispensable collection.)

Dole, S. H. (1964). *Habitable Planets for Man*. New York: Blaisdell (2nd edn, New York: American Elsevier, 1970). (A painstaking and sophisticated discussion.)

Drake, F. D. (1962). *Intelligent Life in Space*. New York and London: Macmillan. (A clearly written, popular account.)

Eddington, A. (1928). *The Nature of the Physical World*. Cambridge: Cambridge University Press. (Chapter VIII, 'Man's place in the universe', is a classic.)

Ehrensvärd, G. (1965). *Man on Another World*. Chicago and London: University of Chicago Press. (See especially Chapter X on 'Advanced consciousness.')

Firsoff, V. A. (1963). *Life Beyond the Earth: A Study in Exobiology*. New York: Basic Books. (A detailed study of the biochemical possibilities for extraterrestrial life.)

Hart, M. H. (1975). An explanation for the absence of extraterrestrials on earth. *Quarterly Journal of the Royal Astronomical Society*, **16**, 128–35. (A perceptive survey of this question.)

Herrmann, J. (1963). *Leben auf anderen Stirnen*. Gütersloh: Bertelsmann Verlag. (A thoughtful and comprehensive survey with special focus on the astronomical issues.)

Hoerner, S. von (1973). Astronomical aspects of interstellar communication, *Astronautica Acta*, **18**, 421–9. (A useful overview of key issues.)

Hoyle, F. (1966). *Of Men and Galaxies*. Seattle: University of Washington Press. (Speculations by one of the leading astrophysicists of the day.)

Huang, S.-S. (1960). Life outside the solar system, *Scientific American*, **202** (4), 55–63. (A useful discussion of some of the astrophysical issues.)

Huygens, C. (1698). *The Celestial Worlds Discovered: New Conjectures Concerning the Planetary Worlds, Their Inhabitants and Productions*. London. (Reprinted by F. Cass & Co., Ltd., London, 1968.) (A classic from another age.)

James, W. (1907). *Pragmatism*. New York: Longmans, Green & Co.

Jeans, J. (1948). Is there life in other worlds? A 1941 Royal Institution lecture reprinted in H. Shapley *et al.* (eds.) (1948), *Readings in the Physical Sciences* (New York: Appleton-Century-Crafts, 1948), pp. 112–117. (A stimulating analysis.)

Kaplan, S. A. (ed.) (1971). *Extraterrestrial Civilization: Problems of Interstellar Communication*. Jerusalem: Israel Program for Scientific Translations. (A collection of Russian scientific papers that present interesting theoretical work.)

Kirk, G. S. & Raven, J. E. (1957). *The Presocratic Philosophers*. Cambridge: Cambridge University Press.

Kuhn, T. (1962). *The Structure of Scientific Revolutions*. Chicago: University of Chicago Press.

Lem, S. (1964). *Summa Technologiae*. Krakow: Wyd. Ltd. (To judge from the ample account given in Kaplan 1971, this book contains an extremely perceptive treatment of theoretical issues regarding extraterrestrial civilizations. I have not, however, been able to consult the book itself.)

MacGowan, R. A. (1962). On the possibilities of the existence of extraterrestrial intelligence. In *Advances in Space Science and Technology*, ed. F. I. Ordway, vol. 4, pp. 39–111. New York and London: Academic Press.

MacGowan, R. A. & Ordway, F. I., III (1966). *Intelligence in the Universe*. Englewood Cliffs: Prentice Hall. (A careful and informative survey of a wide range of relevant issues.)

Nozick, R. (1972). R.S.V.P. – A Story, *Commentary*, **53**, 66–8. (Perhaps letting aliens know about us is just too dangerous.)

Pucetti, R. (1969). *Persons*. New York: Herder and Herder. (A most stimulating philosophical treatment.)

Rood, R. T. & Trefil, J. S. (1981). *Are we Alone?: The Possibility of Extraterrestrial Civilizations*. New York: Scribners.

Sagan, C. (1973). *The Cosmic Connection*. New York: Doubleday. (A well-written, popularly oriented account.)

— (1980). *Cosmos*. New York: Random House. (A modern classic.)

Shapley, H. (1958). *Of Stars and Men*. Boston: Beacon Press. (See especially the chapter entitled 'An inquiry concerning other worlds.')

Shklovskii, I. S. & Sagan, C. (1966). *Intelligent Life in the Universe*. San Francisco, London, Amsterdam: Holden-Day, Inc. (A well-informed and provocative survey of the issues.)

Simmel, G. (1895). Ueber eine Beziehung der Selektionslehre zur Erkenntnistheorie. *Archiv für systematische Philosophie und Soziologie*, **1**, 34–45.

Simpson, G. G. (1964). The nonprevalence of humanoids. *Science*, **143**, 769–75. [= Simpson, G. G. (1964), in *This View of Life: The World of an Evolutionist*, Chapter 13. New York: Harcourt Brace.] (An insightful account of the contingencies of evolutionary development by a master of the subject.)

Strawson, P. F. (1959). *Individuals*. London: Methuen.

Sullivan, W. (1964). *We Are Not Alone*. New York: McGraw-Hill. Revised edn 1965.) (A very well-written survey of the historical background and of the scientific issues.)

Wilkinson, D. H. (1977). *The Quarks and Captain Ahab or: The Universe as Artefact*. Schiff Memorial Lecture. Stanford: Stanford University Press.

Why intelligent aliens will be intelligible

MARVIN MINSKY

When first we meet those aliens in outer space, shall we and they be able to converse? I shall try to show that, yes, we shall – provided they are motivated to cooperate – because we shall both think in similar ways. I shall propose two kinds of arguments for why those aliens may think like us, in spite of their having very different origins. These arguments are based on the idea that all intelligent problem-solvers are subject to the same ultimate constraints – limitations on space, time, and materials. In order for animals to evolve powerful ways to deal with such constraints, they must be able to represent the situations they face, and they must have processes for manipulating those representations.

> **Economics:** Every intelligence must develop symbol-systems for representing objects, causes and goals, and for formulating and remembering the procedures it develops for achieving those goals.
>
> **Sparseness:** Every evolving intelligence will eventually encounter certain very special ideas – e.g. about arithmetic, causal reasoning, and economics – because these particular ideas are very much simpler than other ideas with similar uses.

The 'economics' argument is based on the fact that the power of a mind depends on how it manages the resources it can use. The concept of 'Thing' is indispensable for managing the resources of Space and the substances which fill it. The concept of 'Goal' is indispensable for managing how we use the Time we have available – both for what we do and for what we think about. These notions will be used by aliens, too, both because they are easy to evolve and because there appear to be no easily evolved alternatives for them.

The 'sparseness' theory tries to make this more precise by showing that almost any evolutionary search will soon find certain schemes which have no easily accessible alternatives – that is, other, different, ideas that can serve the same purposes. These ideas or processes seem to be peculiarly isolated, in the sense that the only things which resemble them are vastly more complicated. I will only discuss the specific example of arithmetic, and conjecture that those other concepts of Objects, Causes, and Goals have this same insular character.

> **Critic:** What if those aliens have evolved so far beyond us that their concerns are unintelligible to us, and their technologies and conceptions have become entirely different from ours?

Then communication may not be feasible. My arguments apply only to those stages of mental evolution in which beings are still concerned with surviving, communicating, and expanding their control of the physical world. Beyond that, we may be unable to sympathize with what they come to regard as important. Yet even then we can hope to communicate with the mental mechanisms they use to keep account of space and time; these could remain as sorts of universal currency.

> **Critic:** How can we be sure that things like plants and stones, or storms and streams, are not intelligent in other ways?

If one cannot say in what respects their 'intelligence' is similar, it makes no sense to use the same word. They certainly do not seem to be good at solving the kinds of problems which challenge our intelligence.

> **Critic:** What's so special about solving problems? Anyway, please define 'intelligence' precisely, so that we'll know what we are discussing.

No. It's not one author's place to tell people how to use a word that they already understand. Let us just use 'intelligence' to mean what people usually mean: the ability to solve hard problems – like how to build spaceships and long-distance communication systems.

> **Critic:** Then, you should at least define what a 'hard' problem is. For instance, we know that human intelligence was involved in building the pyramids – yet coral reef polyps build things on even larger scales. Would you claim that we should therefore be able to communicate with them?

No. Humans do indeed solve such problems, but it is only an illusion that coral polyps do. For, there is an important factor of Speed. No single bird discovers how to fly: evolution used a trillion bird-years to find out how – yet, merely tens of human-years sufficed for humans to succeed. And where a person might take several years to find a way to build a structure like an oriole's nest or a beaver's dam, no oriole or beaver could ever learn to do such things without exploiting the ancient nest-machines their genes construct inside their brains. A distinctive aspect of what we call intelligence is this ability to solve very wide ranges of new, different kinds of problems. This is why it makes sense to try to communicate with an individual animal which can learn quickly to solve new, hard problems. But it does not make sense to communicate with the process through which an entire animal species learns to solve problems on time scales which are a million times slower.

Then what enables us to solve hard problems so quickly? Here are some ingredients which seem to me to be so essential that I would expect intelligent aliens to use them, too.

Subgoals—to break hard problem into simpler ones.

Sub-objects—to make descriptions based on parts and relations.

Cause-symbols—to explain and understand how things change.

Memories—to accumulate experience about similar problems.

Economics—to allocate scarce resources efficiently.

Planning—to organize work, before filling in details.

Self-awareness—to provide for the problem-solver's own welfare.

Still, one might ask, aren't these only a few of myriads of possibilities? Why cannot our aliens do all such things in completely alien ways? To answer this, I shall show that these problem-solving schemes are not as arbitrary as they seem.

The sparseness principle

Why does it seem so obvious to us that Two and Two must equal Four? Such mysteries – why certain concepts seem to come into our minds as though they need no prior experience or evidence – have long concerned philosophers. My answer is that this may be due, at least in part, to the following 'computational phenomenon.'

The Sparseness Principle: Whenever two relatively simple processes have products which are similar, those products are likely to be completely identical!

Because of this, we can expect certain '*a priori*' structures to appear, almost always, whenever a computation system evolves by selection from a universe of possible processes. The ideas of number and arithmetic are examples of this, and my conjecture is that this may be why different people can communicate so perfectly about such matters, although their minds may differ in many other ways. And so, it may apply to aliens, too. I will explain the sparseness principle by recounting two anecdotes. One involves a mathematical experiment, the other, a real-life episode.

A mathematical experiment

I once set out to explore the behaviors of all possible processes – that is, of all possible computers and their programs. There is an easy way to begin that search: one just begins to list possible finite sets of rules, one by one. That is easy to do, by using a method which Alan Turing described in 1936; these are what, today, we call 'Turing machines'. Naturally, I did not get very far, because the variety of such processes grows exponentially with the number of rules in each set. However, with the help of my student Daniel Bobrow, we managed to examine the first few thousands of such machines – and we found that among them all there were only a few distinct kinds of behaviors. Some of them simply stopped, without accomplishing anything. Many of the others just erased their input data and did nothing else. Most of the remainder quickly got trapped in circles, senselessly repeating the same steps over again. There were only a few left which did anything interesting at all – and these were all essentially the same: each of them performed a 'counting operation' that repeatedly increased by one the length of a string of symbols. In honor of their ability to do what resembles a fragment of simple arithmetic, let us call these 'A-Machines'. Let us think of this exploration as exposing parts of some infinite 'universe of possible computational structures'. Then this tiny fragment of evidence suggests that such a universe may look something like this.

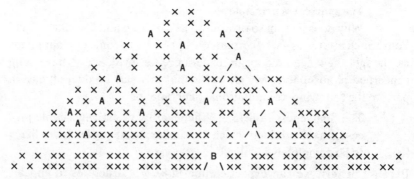

The 'X's represent those useless processes which do scarcely anything at all, while the 'A's represent those little 'counting machines' – which, in effect, are all identical! Little processes like these, inside our minds, could be seeds of our more mature ideas about arithmetic. My point is that it seems inevitable that, somewhere, in a growing mind some A-machines must come to be. Now, possibly, there are some other, really different ways to count. So there may appear, much, much later, some of what we represent as 'B-machines' – which are processes that act in ways which are similar, but not identical to, how the A-machines behave. But, our experiment hints that even the very simplest possible B-machine will be so much more complicated that it is unlikely that any brain would discover one before it first found many A-machines.

I think of this little thought-experiment as resembling an abstract version of those first experiments in which Stanley Miller and Harold Urey set out to explore, with real chemicals, the simplest combinations of constituents. They started with a few elements like hydrogen, oxygen, nitrogen, carbon, and phosphorus and found that those chemicals reacted first to make simple molecules and then went on to form peptides, sugars, nucleotides and what-not. Of course, we would have to wait much, much, longer before the appearance of tigers, woodpeckers, or Andromedans.

A real-life episode

Once, when still a child in school, I heard that 'minus times minus is plus.' How strange it seemed that negatives could 'cancel out' – as though two wrongs might make a right, or that 'this statement lies' could be a truth. I wondered if there could be something else, still like arithmetic, except for having yet another 'sign'. Why not make up some number-things, I thought, which go not just two ways, but three? I searched for days, making up new little multiplication tables. Alas, each system ended either with impossible arithmetic (e.g. with One and Two the same), or something with no signs at all, or something with an extra sign. I gave up, eventually. If I had had the courage to persist, as Gauss did, I might have discovered the arithmetic of complex numbers or, as Pauli did, the arithmetic of spin matrices. But no one ever finds a three-signed imitation of arithmetic – because, it seems, they simply do not exist.

Try, yourself, to make a new number system that is like the ordinary one, except that it 'skips' some number, say, 4. It just won't work. Everything will go wrong. For example you will have to decide what is '2 plus 2'. If you say that this is 5, then 5 will have to be an even number, and so also must 7 and 9. And then, what is 5 plus 5? Should it be 8, or 9 or 10? You will find that, to make the new system be at all like arithmetic, you will have to change the properties of all the other numbers. Then, when you are finished, you will find that you have only changed those numbers' names, and not their properties at all.

Similarly, you could try to make two different numbers be the same – say, 139 and 145. But, then, to make subtraction work, you will have to make 6 be the same as zero, and 4 plus 5 will then be 3. Suddenly, you find that the sum of two positive numbers is smaller than either of them – and that scarcely resembles arithmetic at all. (In fact, this leads to 'modular arithmetic', which has a certain usefulness in abstract mathematics – but is worse than useless for keeping track of real things.) And so it goes.

There is just no way to take a single number out, or put another in. Nor can you change a single product, sum, or prime. What gives arithmetic this stark and singular rigidity? One cannot make the smallest hole in it, or make it stretch or bend the slightest bit. You have to take it as it stands, the whole thing, all or none, unchangeable – because it is isolated as an island in that universe of processes. That self-same A-machine exists, immutably complete, as part of every other process which can generate an endless chain of different things.

I sometimes wonder if it is dangerous to make our children dwell so long upon arithmetic – since, when it is seen this way, it leads to such a singularly barren world. True, some children find in it a universe of different things to do. Most children, though, just find it dull, a source of endless rote and pointless pain; it is like the tedium of working clay too cold to mold to any other shape.

From all this, I conclude that any entity who searches through the simplest processes will soon find fragments which do not merely resemble arithmetic but *are* arithmetic. It is not a matter of inventiveness or imagination, only a fact about the geography of the universe of computation, a world far more constrained than that of real things.

> **Thesis:** All processes or formalisms which resemble arithmetic are either identical to it, or else unthinkably complicated. This is why we can communicate perfectly about numbers.

What has this to do with aliens? Only that they too must have evolved by searching through some universe of possible processes – and any evolutionary process must first consider relatively simple systems, and thus discover the same, isolated, islands of efficiency.

Finally, we ought to ask why processes occur that way – without some similar ones nearby. It is hard to formulate this precisely, because the meaning of 'similar' depends on what you want to use it for. One way to explain it is to point out that a small set of rules can generate a vast world of implications and consequences. But there is no converse, because, usually, a large and complex thing cannot be described by a small set of rules. This is simply a matter of arithmetic: there just are not enough small sets of rules to go around! And that explains why we cannot take some set of rules, use them to generate a universe of consequences, make a few changes in that universe, and then describe that end result, again in terms of only a few rules – for, now, that altered universe is one of miracles, and not of laws. There are not enough small sets of rules to produce the effect of continuity.

Causes and clauses

An alien mind would probably be entirely different from ours – if how we think were just an evolutionary accident. And then, communication would probably not be feasible. But although every evolution is composed of many accidents, still, each evolution tends first to try relatively simple ways at every stage. Since we are the first on earth to develop complex languages and, since these are likely to employ very many, relatively simple principles, it is possible that other alien species will share many of these. I shall propose this in a form so strong that it may seem entirely preposterous at first: I shall argue that many aspects of our language-grammar forms may be almost inescapable.

Why do we say things like '*It* soon will start to rain?' Why must we always postulate some agent-cause, no matter that there is no actor on the scene? No matter if we are right or wrong, we will find one, or imagine one. And, in the mind, I claim we seek some Cause for every Difference, Move, or Change. Sometimes this is forced on us by our language-syntax – but I claim this is not merely a matter of verbal form, but stems from deeper causes in the ways we think. My guess is that, even before our ancestors began to speak, they first developed special brain-machinery for representing objects, differences, and causes, and these were later reflected in our language-grammars. Specifically, I suspect that many of our thought-processes are based on using the following kinds of mental symbol-representations:

> *Object-symbols* represent things, ideas, or processes. In languages, they often correspond to Nouns. Our minds tend to describe every situation, real or mental, in terms of separate object-things and relations between them.
>
> *Difference-symbols* represent differences between, or changes in, *Objects*. In languages, they often correspond to Verbs. When any object undergoes a change, or two objects are compared, the mind ascribes some *Differences* to them.
>
> *Cause-symbols*. When any *Difference* is conceived, the mind is made to find a *Cause* for it – something which is held responsible.

And we use a clever mental trick to represent causes in much the same ways that we represent objects.

Clause-structures. For describing complicated situations, we have a trick which lets us treat any expression or description, however complicated, as though it were a single component of another description. In languages, this corresponds to using embedded phrases and clauses.

It is that final, self-embedding trick – of representing prior thoughts as things, which gives our minds their awesome power. For this permits us to re-use the same brain-machinery over and over again, at each step replacing an entire conceptualization by a compact symbol. That way, we can build up gigantic structures of ideas, as easily as our children build great bridges and towers from simple, separate blocks. That way, we can build new ideas from old ones – and that is what enables us to think. The same is true of our computers.

This must be why our languages, too, have structures which we can re-use: our thoughts, themselves, must use the same machinery repeatedly. That is how our thoughts extend themselves to infinite variety. And, unless aliens do that too, they cannot turn their thoughts upon the prior products of their thoughts. Without this trick of turning symbols on themselves, one cannot have general intelligence – however excellent may be one's repertoires of other skills.

> **Critic:** You might as well argue that the aliens will speak English, if you claim they, too, use nouns and verbs and compound, embedded sentences. But what if they don't think in terms of objects and actions at all?

I do not think that's an accident – the way we think in terms of thing and cause. Whatever may occur, that way of representing it leads us always to wonder who or what is responsible. And so, this evolutionary trick leads us to search to find dependencies which help to predict – *and hence to control* – not just the world outside but also what may happen in the mind. Perhaps it is also why we all grow up believing in a Self: perhaps that 'I' – in 'I just had a good idea' – stems from that same machinery. For if you are compelled to find a thing to cause the things you do – why, then, that something needs a name. You call it 'me'. I call it 'you'.

> **Critic:** But what's so great about dependencies? Why can't some aliens perceive entire scenes as wholes instead of breaking them down into those clumsy parts? Why not, instead, see what there really is, holistically, – a steady flow of flux in space in time,

instead of arbitrary form-filled, mind-made fragments of approximations to reality?

It surely is a healthy tendency to yearn for better ways to see the world. But worshipping yet undiscovered transcendental schemes can blind us to the power we draw from our usual ways of separating things. Each animal must pay a corresponding price, in energy and nourishment, for each machine it carries in its brain. Just as clause-structure in language lets us focus our entire word-machine on each part of a description, our concept of seeing separate things lets us divide situations into parts, and then apply our whole mind-machine to each part of the problem. Enthusiasts of 'holism' have never understood the hidden cost a mind would have to pay in order to 'see everything at once'. We would never be able to see anything clearly at all.

There have been many speculations on how brains might use something like holograms for memories. But, on one side, there is no evidence for this. On the other, there would be few advantages to it. Holograms store no more information than other methods, and computer scientists know other, better ways to add redundancy to memory to make it robust and injury resistant. It is true that holograms can simplify certain kinds of recognitions; for instance, deciding whether a picture contains copies of some specific other picture. But that also makes it much more difficult to make most other kinds of decisions, e.g. to say whether a picture contains two subpictures that share some specified relationship. In fact, a hologram may be almost the worst possible way to represent relations among the things it represents, because memory and learning are only useful when they inform us about relations which are at least partially predictable. We do not want our memories to give equal weights to every arbitrary feature of situations. If a scene contains 50 features, one does not want equally to consider all the quadrillion possible subsets of those features. And so we need some methods for isolating and grouping which can emphasize the most usefully predictable subsets. In short, without the additional constraints on relations between features which result in the concept of an 'object', we would simply never see the same thing twice. And then we would have no way to learn from our experience. No knowledge could accumulate.

Causes and goals

How does having memory help when no two problems are ever quite the same in all respects? Our past experience would seem to have no relevance, unless we had some ways to see which aspects of the world remain the same, while the other aspects change. This is why knowledge cannot have much use unless expressed in terms of relations between 'predictable' features and the actions which we can take. But, given these, it then may become possible to predict which actions might cause undesirable features to disappear.

To say that 'Y happened because of an X' is, in effect, to say that X can help one to predict which actions can lead to Y. An animal can help to control its environment, if it can find such 'causes' – fragments of predictability that work better than chance. But such predictions are not useful when too many small effects 'add up'. For, what are causes, anyway? The very concept of a cause involves a certain element of style: a causal explanation must be brief! Unless an explanation is compact, we cannot use it to predict! We would agree that X is a cause of Y, if we see that Y depends much more on X than on most other things. But we would not call X a cause if X were a discourse which carries on and on, mentioning everything else in the whole world and never comes to any 'point'. This applies to objects, too.

> There cannot be any objects, things, or causes in worlds where everything that happens depends, more or less equally, on everything else that happens.

It makes no sense to talk about a 'thing' in such a world, because our very notion of a 'thing' assumes some constellation of properties which stay the same or change in ways which we can predict, when other things around it change. When you move an object, its location will be changed – but not its color, weight, material, size or shape. How convenient that our world lets us change a thing's place, and still leave so many other properties unchanged! If this were not the case, the number of combinations of which we had to keep track would grow exponentially with the number of features we perceive, and we would have no sense of what 'causes' things to happen.

> To deal with something complicated, one must find a way to describe it in terms of substructures within which the effects of actions tend to be localized. To know the cause of a phenomenon is to know, at least in principle, what can change or control it without changing everything else. This is useful when it enables us to change one thing without making other things 'worse'.

In order for a mind to discover causes in its world, it must have sensors which detect changes which are predictably related to the actions it can take. Fortunately, evolution tends automatically to select just such matched sets of sensors and effectors, because, in virtually any environment, an animal's survival will be enhanced if its actions are based on good predictions. So, we can expect evolutionary processes to accumulate mechanisms which reflect the causal laws that operate in their environments! And it would seem that the most powerful methods are those which let one make predictions about the effects of contemplated chains of action – that is, the ability to make plans.

Problems seem hard when it isn't obvious what to do! The most general way we know to solve problems is to set up systems which have some way to make 'progress toward a goal'. In the late 1950s, A. Newell and H. A. Simon worked out a theory of what they called the 'General Problem Solver' – a theory of how to reach a goal by 'making progress' by finding actions which can replace each problem that has a 'high-level of difficulty' by other problems which each have lower-levels of difficulties. Now no one can prove that all intelligent problem-solvers, however alien, must use this selfsame principle. But until we find another, comparably general idea – and none is on our horizon – it is hard to imagine how an intelligence could evolve without exploiting some such idea of goal.

Reliable communication

Before we ask how aliens communicate, we ought to ask how humans can do so. Is ever there a word which means the same to any two of us? Everyone must have wondered once, 'could two persons have different meanings for every word, yet never sense that anything was wrong?' What if each thing that is green (or blue) to me were blue (and green) to you? The Sparseness theory claims that we need have no fear of that, at least for technical concepts, since one of the two outwardly indistinguishable meanings would probably be vastly more complicated than the other, and never have been conceived in the first place. Sparseness means we can trust one another.

We know very little of where that idea might lead, because we know so little about how sparseness isolates any particular concepts. But the general idea does seem to support the mathematical and physical intuitions proposed by Hans Freudenthal (1960; see Freudenthal, this volume) in *LINCOS*, his book on alien communication – perhaps even in regard to the miniature models he suggests for discussing social and administrative subjects. There is one problem, though: introspection is a poor guide for guessing which of our common-sense concepts are really 'simple', because many things we find easy to do use brain-machines whose complexity we cannot sense. For example, we find it easy to stand on two feet – but some aliens might find that quite astonishing.

What other ideas are likely to be 'universal' – in the sense of being islands in that sea of possible ideas? Surely the mathematical notions of utility, linear approximation, probability, and the simplest program-like processes. These could serve to communicate much about trade and commerce, basic facts of biology, and even many principles of mental life, e.g. about objects, goals, and memories. At some point, though, sparseness must fail, for things which are more complicated will have all kinds of variations and alternatives, and communication will encounter obstacles of every sort. So, I see little more to say of this today, with any scientific certitude. Tomorrow, there could be more, perhaps because of soon-to-come gains in computational power, which could let us explore a little further into the mysterious ocean of all possible simple machines. There, we might find a few more ideas, isolated enough to share with other minds. Such explorations, too, might tell us more about the origin of life itself, by showing us the simplest schemes which could support the simplest forms of evolutionary search.

References

Ernst, G. & Newell, A. (1969). *GPS: A case study in generality and problem-solving*. New York: Academic Press. (Shows how many different kinds of problems can be solved by mechanisms based on recognizing differences and trying to remove them.)

Freudenthal, H. (1960). *LINCOS: Design of a Language for Cosmic Intercourse*. Amsterdam: North-Holland. (Lincos drafts scenarios for communicating with aliens, starting with elementary arithmetic and then working up to many concepts of physics, psychology, and social behavior. Freudenthal's profound constructions might serve as a basis for exploring the distribution of important simple concepts.)

Lenat, D. (1982). The Nature of Heuristics. *Artificial Intelligence*, **19**, 189–249. (Lenat's computer program discovered many principles of arithmetic in the course of an evolutionary search.)

Turing, A. (1937). On computable numbers, with an application to the Entscheidungsproblem. *Proceedings of the London Mathematical Society*, **42** (2), pp. 230–65. (Reprinted in Martin Davis (ed.) (1965); *The Undecidable*, New York: Raven Press.)

PART 4

Where are they?

The question is attributed to Enrico Fermi, who asked it at a Los Alamos dinner party during the Manhattan Project. If extraterrestrials really existed, he wanted to know, then 'where are they?'

This by now famous question, about which at least one entire book has been written (Hart & Zuckerman 1982), has been taken with varying amounts of seriousness by different parties to the ETI debate. Critics tend to take it very seriously, while proponents tend not to. The logic behind the question is expressed as follows by Jill Tarter (who denies, however, that the argument is valid):

'(1) *If* extraterrestrial civilizations have existed elsewhere and "elsewhen" in our galaxy,

(2) *and if* interstellar travel/colonization/migration is inevitable for at least one of them,

(3) *then* simple calculations indicate that an expanding wave of colonization will fill the galaxy on a time scale short compared to the lifetime of the galaxy,

(4) *but* we do not "see" them here.

(5) *Therefore*, (1) is wrong; there has never been another technological civilization anywhere or "anywhen" in our galaxy except the earth.'

(Tarter 1985)

Michael Hart advanced essentially this argument in his influential paper 'An explanation for the absence of extraterrestrials on earth' (Hart 1975). His explanation was that there *are* no extraterrestrials; in other words, proposition (5) is true. Proponents of ETI, on the other hand, claim that absence of extraterrestrials here does not mean there are not any elsewhere. Frank Drake, for example, explains the absence of aliens on earth by saying that they have no reason to make the journey from their own planet to here. According to Drake, the power required to make an interstellar journey is so large that residents of one star system have no economic incentive to leave it. The answer to the 'Where are they?' question, he says, is that they 'are living comfortably and well in the environs of their own star' (Drake 1980).

But even if this were true, a question would still remain: Where are their artifacts? For although the aliens themselves might remain on their home planet, a plausible argument can be given that their instrumental probes ought to be here by now. In 1980 just such an argument was given by Tulane University theoretical physicist Frank Tipler. Originally published in the *Quarterly Journal of the Royal Astronomical Society* (*QJRAS*) his article, 'Extraterrestrial intelligent beings do not exist,' is reprinted here.

Tipler's argument is built upon the idea of von Neumann machines, robotic creatures capable of self-reproduction. The machines were originally envisioned by mathematician John von Neumann who called them self-reproducing automata; Tipler calls them self-reproducing universal constructors, machines capable of making any device, including a copy of itself. According to Tipler, a sufficiently advanced technical society would have made such machines and would have sent them out to fill the galaxy. But since the aliens' von Neumann machines aren't here, it follows that the aliens do not exist.

Since it reasons from what is a bare abstract possibility – von Neumann machines – Tipler's argument may seem implausible. The fact is, however, that Tipler's critics do not challenge the idea of self-reproducing automata. The reason for this may be that such machines are essentially what DNA molecules are. When von Neumann outlined his ideas in a 1948 lecture entitled 'The general and logical theory of automata,' he said that such a device would have four components: A, an automatic factory; B, a duplicator; C, a controller; D, an instruction program. According to Freeman Dyson, 'Five years later Crick and Watson discovered the structure of DNA, and now every child learns in high school the biological identification of von Neumann's four components. D is the genetic materials, RNA and DNA; A is the ribosomes; B is the enzymes RNA and DNA polymerase; and C is the repressor and derepressor control molecules . . . . So far as we know, the basic design of every microorganism larger than a virus is precisely as von Neumann said it should be' (Dyson 1979).

Tipler describes how and why an advanced extraterrestrial culture would send an army of von Neumann machines out into the cosmos.

Contrary to what Drake has said about the expense of colonizing other star systems, Tipler calculates that for a technologically well-developed civilization, 'the exploration of the galaxy would cost about 3 billion dollars, about one-tenth the cost of the *Apollo* program.' Assuming then that von Neumann machines are in principle possible, and that with them a suitably advanced race of extraterrestrials could have covered the galaxy for a small expenditure, the questions return: Where are they? Why aren't they here?

Carl Sagan and William I. Newman replied to Tipler in their article 'The solipsist approach to extraterrestrial intelligence,' published in *QJRAS* in 1982, and reprinted in this volume. Author of the acclaimed *Cosmos*, and Professor of Astronomy and Space Sciences at Cornell University, Sagan is undoubtedly the world's best known SETI advocate. William I. Newman, a former student of Sagan's at Cornell, is Associate Professor of Earth and Space Sciences at the University of California, Los Angeles.

Sagan and Newman advance three alternatives to Tipler's claim that extraterrestrials do not exist. To begin, they question Tipler's scenario of a galaxy dominated by von Neumann machines. For these 'implacable replicators' will either have or lack a built-in principle of reproductive restraint. But if their reproduction is limited, this would explain why such machines are not found on earth. If, on the other hand, such machines would reproduce endlessly, then they would threaten even the culture which contemplates making them, and thus they would not be made to begin with. In either case, the absence of such machines on earth is satisfactorily explained without denying the existence of intelligent aliens.

A second possibility is that the migration wave of an interstellar culture would take longer to cross the galaxy than Tipler allows. And a third case is that cultures who are aggressive enough to plan galactic colonization will destroy themselves before they get very far.

The 'Where are they?' dispute won't be settled, perhaps, until we actually 'do the experiment,' until we listen systematically for their signals. The next section describes what such an experiment would be like.

References

Drake, F. D. (1980). *N* is neither very small nor very large. *Strategies for the Search for Life in the Universe*. ed. M. D. Papagiannis, pp. 27–34. Dordrecht: D. Reidel.
Dyson, F. (1979). *Disturbing the Universe*. New York: Harper & Row.

Hart, M. (1975). An explanation for the absence of extraterrestrials on Earth. *Quarterly Journal of the Royal Astronomical Society*, 16, 128–35.

Hart, M. H. & Zuckerman, B. (eds.) (1982). *Extraterrestrials: Where are They?* New York: Pergamon Press.

Tarter, J. (1985). Planned observational strategy for NASA's first systematic search for extraterrestrial intelligence. In *Interstellar Migration and the Human Experience*, ed. E. M. Jones & B. R. Finney. Los Angeles: University of California Press, in press.

von Neumann, J. (1948). The general and logical theory of automata. In *Collected Works*, vol. 5, ed. A. H. Taub, pp. 288–328. New York: Macmillan, 1961–3. See also *John von Neumann: Theory of Self-Reproducing Automata*, ed. and completed by A. W. Burks. Urbana, Ill.: University of Illinois Press, 1966.

Extraterrestrial intelligent beings do not exist*

FRANK J. TIPLER

Summary
It is argued that if extraterrestrial intelligent beings exist, then their spaceships must already be present in our solar system.

1   Introduction to the argument

One of the most interesting scientific questions is whether or not extraterrestrial intelligent beings exist. This question is not new; in one form or another it has been debated for thousands of years (*1*). The contemporary advocates for the existence of such beings seem to be primarily astronomers and physicists, such as Sagan (*2*), Drake (*3*), and Morrison (*4*), while most leading experts in evolutionary biology, such as Dobzhansky (*5*), Simpson (*6*), Francois (*7*), Dobzhansky *et al.* (*8*), and Mayr (*9*), contend that the Earth is probably unique in harbouring intelligence, at least amongst the planets of our Galaxy. The biologists argue that the number of evolutionary pathways leading from one-celled organisms to intelligent beings is minuscule when compared with the total number of evolutionary pathways, and thus even if we grant the existence of life on 10^9 to 10^{10} planets in our Galaxy, the probability that intelligence has arisen in our Galaxy on any planet but our own is still very small. I agree with the biologists; I shall argue in this paper that the probability of the evolution of creatures with the technological capability of interstellar communication within five billion years after the development of life on an Earth-like planet is less than 10^{-10}, and thus we are the only intelligent species now existing in this Galaxy. The basic idea of my argument is straightforward and indeed has led other authors, such as Fermi (*10*), Dyson (*11*), Hart (*12*), Simpson (*6*), and Kuiper & Morris (*13*), to

* Research supported by the National Science Foundation under grants number MCS-76-21525 and PHY-77-15191.

conclude that extraterrestrial intelligent beings do not exist: if they did exist and possessed the technology for interstellar communication, they would also have developed interstellar travel and thus would already be present in our solar system. Since they are not here (*14,15*), it follows that they do not exist. Although this argument has been expressed before, its force does not seem to have been appreciated. I shall try to rectify this situation by showing that an intelligent species with the technology for interstellar communication would necessarily develop the technology for interstellar travel, and this would automatically lead to the exploration and/or colonization of the Galaxy in less than 300 million years.

To begin with, we must assume that any intelligent species which develops the technology for interstellar communication must also have (or will develop in a few centuries) technology which is at least comparable to our present-day technology in other fields, particularly rocketry. This is actually a consequence of the principle of mediocrity (*16*) (that our own evolution is typical), which is usually invoked in analyses of interstellar communication. However, this assumption is also an essential one to make if interstellar communication via radio is to be regarded as likely. If we do not assume that an advanced species knows at least what we know, then we have no reason to believe an advanced species would use radio waves, for they may never have discovered such things. In the case of rocket technology, the human species developed rockets some 600 years before it was even aware of the existence of radio waves, and present-day chemical rockets can be regarded as logical extensions of early rocket technology.

In addition to a rocket technology comparable to our own, it seems likely that a species engaging in interstellar communication would possess a fairly sophisticated computer technology. In fact, Sagan has asserted (*17*) that 'Communication with extraterrestrial intelligence . . . will require . . . , if our experience in radio astronomy is any guide, computer-actuated machines with abilities approaching what we might call intelligence.' Furthermore, the Cyclops (*18*) and SETI (*19*) proposals for radio telescopes to search for artificial extraterrestrial radio signals have required some fairly advanced data-processing equipment. I shall therefore assume that any species engaging in interstellar communication will have a computer technology which is not only comparable to our present-day technology, but which is comparable to the level of technology which we know is possible, which we are now spending billions of dollars a year to develop, and which a majority of computer experts believe we will actually possess within a century. That is, I shall assume that such a species will eventually develop a self-replicating universal constructor with intelligence comparable to our present-day technology, but which is comparable to the human level – such a machine should be developed within a century, according to the experts (*20,21,22*) – and such a machine combined with present-day rocket technology, would make it possible to explore and/or colonize the Galaxy in less than 300 million years, for an initial investment less than the cost of operating a 10 MW microwave beacon for several hundred years, as proposed in SETI (*19*). It is a deficiency in computer technology, not rocket technology, which prevents us from beginning the exploration of the Galaxy tomorrow.

2   **The general theory of space exploration and colonization**
    In space exploration (or colonization), one chooses a strategy which maximizes the probable rate of information gained (or regions colonized) and minimizes the cost of the information, subject to the constraints imposed by the level of technology. Costs may be minimized in two ways: first, 'off-the-shelf' technology is to be used as far as possible to reduce the research and development costs; second, resources which could be used for no other purpose should be utilized as far as possible. The resources available in uninhabited stellar systems cannot be utilized for any human purpose unless a space vehicle is first sent; therefore, any optimal exploration strategy must utilize the material available in the other stellar system as far as possible. With present-day technology, such utilization could not be very extensive, but with the level of computer technology assumed in the previous section, these otherwise useless resources can be made to pay for virtually the entire cost of the exploration program.
    What one needs is a self-reproducing universal constructor, which is a machine capable of making any device, given the construction materials and a construction program. In particular, it is capable of making a copy of itself. Von Neumann has shown (*23*, cf. *24*) that such a machine is theoretically possible, and in fact a human being is a universal constructor specialized to perform on the surface of the Earth. (Thus the manned space exploration [and colonization] program outlined in (*11,12,13*) is just a special case of the exploration strategy discussed below.)
    The payload of a probe to another stellar system would be a self-reproducing universal constructor with human level intelligence (hereafter called a von Neumann machine) together with an engine for slowing down once the other stellar system is reached, and an engine for travelling from one place to another within the target stellar system – the latter could be an electric propulsion system (*25*), or a solar sail (*26*). This machine would be instructed to search out construction material with which to

make several copies of itself and the original probe rocket engines. Judging from observations of our own solar system (*27*), what observations we have of other stellar systems (*28*), and the vast majority of contemporary solar system formation theories (*29*), such materials should be readily available in virtually any stellar system – including binary star systems – in the form of meteors, asteroids, comets, and other débris from the formation of the stellar system. Whatever elements are necessary to reproduce the von Neumann machine, they should be available from one source or another. For example, the material in the asteroids is highly differentiated; many asteroids are largely nickel–iron, while others contain large amounts of hydrocarbons (*27*).

As the copies of the space probe were made, they would be launched at the stars nearest the target star. When these probes reached these stars, the process would be repeated, and so on until the probes had covered all the stars of the Galaxy. Once a sufficient number of copies had been made, the von Neumann machine would be programmed to explore the stellar system in which it finds itself, and relay the information gained back to the original solar system from which the exploration began. In addition, the von Neumann machine could be programmed to use the resources of the stellar system to conduct scientific research which would be too expensive to conduct in the original solar system.

It would also be possible to use the von Neumann machine to colonize the stellar system. Even if there were no planets in the stellar system – the system could be a binary star system with asteroid-like débris – the von Neumann machine could be programmed to turn some of this material into an O'Neill colony (*30*). As to getting the inhabitants for the colony, it should be recalled that all the information needed to manufacture a human being is contained in the genes of a single human cell. Thus if an extraterrestrial intelligent species possessed the knowledge to synthesize a living cell – and some experts assert (*31,32*) the human race could develop such knowledge within 30 years – they could program a von Neumann machine to synthesize a fertilized egg cell of their species. If they also possessed artificial womb technology – and such technology is in the beginning stages of being developed on Earth (*33*) – then they could program the von Neumann machine to synthesize members of their species in the other stellar system. As suggested by Eiseley (*34*), these beings could be raised to adulthood by robots in the O'Neill colony, after which they would be free to develop their own civilization in the other stellar system.

Suggestions have occasionally been made (*35*) that other solar systems could be colonized by sending frozen cells via space probes to the stars.

However, it has not yet been shown (*36–39*) that such cells would remain viable over the long periods required to cross interstellar distances. This difficulty does not exist in the above-outlined colonization strategy; the computer memory of the von Neumann machine can be made so that it is essentially stable over long periods of time. If it is felt that the information required to synthesize an egg cell would tax the memory storage space of the original space probe, the information could be transmitted via microwave to the von Neumann machine once it has had time to construct additional storage capacity in the other solar system. The key point is that, once a von Neumann machine has been sent to another solar system, the entire resources of that solar system become available to the intelligent species which controls the von Neumann machine; all sorts of otherwise-too-expensive projects become possible. It would even be possible to program the von Neumann machine to construct a very powerful radio beacon with which to signal other intelligent species!

Thus the problem of interstellar travel has been reduced to the problem of transporting a von Neumann machine to another stellar system. This can be done even with present-day rocket technology. For example, Hunter (*40*, cf. *41*) has pointed out that by using a Jupiter swingby to approach the Sun and then giving a velocity boost at perihelion, a solar system escape velocity v_{es} of about 90 km s^{-1} ($\sim 3 \times 10^{-4} c$) is possible with present-day chemical rockets, even assuming the launch is made from the surface of the Earth. As discussed in references (*28*) and (*29*), most other stars should have planets (or companion stars) with characteristics sufficiently close to those of the Jupiter–Sun system to use this launch strategy in reverse to slow down in the other solar system. The mass ratio μ (the ratio of the payload mass to the initial launch mass) for the initial acceleration would be 10^3, so the total trip would require $\mu < 10^6$ (less than, since the 10^3 number assumed an Earth-surface launch), quite high, but still feasible. With Jupiter swingby only, the escape velocity would be $\sim 1.5 \times 10^{-4} c$ with $\mu = 10^3$. The *Voyager* spacecraft will have (*42*) a solar escape velocity of about $0.6 \times 10^{-4} c$ with $\mu = 850$.

It thus seems reasonable to assume that any intelligent species would develop at least the rocket technology capable of a one-way trip with deceleration at the other stellar system, and with a travel velocity v_{es} of $3 \times 10^{-4} c$. At this velocity, the travel time to the nearest stars would be between 10^4 and 10^5 years. This long travel time would necessitate a highly developed self-repair capacity, but this should be possible with the level of computer technology assumed for the payload (*43*). Nuclear power-sources could be developed which would supply power for that length of time. However, nuclear power is not really necessary. If power

138   *Frank J. Tipler*

utilization during the free-fall period was very low, even chemical reactions could be used to supply the power. As v_{es} is of the same order as the stellar random motion velocities, very sensitive guidance would be required, but this does not seem to be an insuperable problem with the assumed level of computer technology.

Because of the very long travel times, it is often argued (*44*) that interstellar probes would be obsolete before they arrived. However, in a fundamental sense a von Neumann machine cannot become obsolete. The von Neumann machine can be instructed by radio to make the latest devices after it arrives at the destination star.

Restricting consideration to present-day rocket technology is probably too conservative. It is likely that an advanced intelligent species would eventually develop rocket technology at least to the limit which we regard as technically feasible today. For example, the nuclear pulse rocket of the Orion Project pictured (*45*) a solar escape velocity v_{es} of $3 \times 10^{-2}\,c$ with $\mu = 36$ for a one way trip and deceleration at the target star. The cost of the probe would be 3×10^{12} in 1979 prices, almost all of the money being for the deuterium fuel. This is about the present GNP of the United States. Project Daedalus (*43*), the interstellar probe study of the British Interplanetary Society, envisaged a stellar fly-by via nuclear pulse rocket (no slow-down at the target star), with $v_{es} = 1.6 \times 10^{-1}\,c$, $\mu = 150$, and a cost of 9×10^{11}. As before, almost all the cost is for the helium-3 fuel (at 1960 prices). With slow-down at the target star, $\mu = 2 \times 10^4$, the cost would be 1.4×10^{14}, or almost 100 times the United States GNP, and it would require centuries to extract the necessary helium-3 from the helium source in the Daedalus study, the Jovian atmosphere.

The cost of such probes is far beyond present-day civilization. However, note that almost all the cost is for the rocket fuel. Building the probe itself and testing it would cost relatively little. A possible interstellar exploration strategy would be to design a probe capable of $v_{es} = 0.1\,c$, record the construction details in a von Neumann machine, launch the machine via a chemical rocket at $3 \times 10^{-4}\,c$ to a nearby stellar system, and program the machine to construct and *fuel* several high velocity ($0.1\,c$) probes with von Neumann payloads in the other system. When these probes reach their target stars, they would be programmed to build high velocity probes, and so on. In this way the investment on interstellar probes by the intelligent species is reduced to a minimum while maximizing the rate at which the Galaxy is explored. (The von Neumann machines could conceivably be programmed to develop the necessary technology in the other system. This would reduce the initial investment even further.) The disadvantage, of course, is the fact that for 10^4 years, there is no information on other stellar systems reaching the original solar system. There is a trade-off between the cost of the first probe and the time interval the intelligent species must wait before receiving any information on other stellar systems. But with second-generation probes with $v_{es} = 0.1\,c$, new solar systems could be explored at the rate of several per year by 10^3 years after the original launch. The intelligent species need only be patient and launch a sufficient number of initial probes at $v_{es} = 3 \times 10^{-4}\,c$ so that at least one succeeds in reproducing itself (or in making a high velocity probe) several times. This number will depend on the failure rate. Project Daedalus *(43)* aimed at a mission failure rate of 10^{-4}, and the designers argued that such a failure rate was feasible with on-board repair. If we adopt this failure rate and assume failures to be statistically independent, then only three probes need be launched to reduce the failure probability to 10^{-12}. Judging by contemporary rocket technology, the cost of the initial low velocity probes would be less than $\$1 \times 10^9$ each, since von Neumann probes would make themselves and the original R & D costs would be small – von Neumann machines would originally be developed for other purposes *(46)*. Thus the exploration of the Galaxy would cost about 3 billion dollars, about one-tenth the cost of the *Apollo* program.

To maximize the speed of exploration and/or colonization, one must minimize $[d_{av}/v_{es} + t_{const}]$, where d_{av} is the average distance between stars and t_{const} is the time needed for the von Neumann machine to reproduce itself and the space probe. The time t_{const} will be much larger for $v_{es} = 0.1\,c$ probes than for $10^{-4}\,c$ probes. I would guess the minimum to be obtained for $v_{es} = 5 \times 10^{-2}\,c$ and $t_{const} = 100$ years. With $d_{av} = 5$ light years, this gives a rate of expansion of 2.5×10^{-2} ly/yr, and thus the Galaxy could be explored in 4 million years. For the purposes of this article I shall assume only present-day rocket technology, which would give an expansion rate of 3×10^{-4} ly/yr, and the exploration of the Galaxy in 3×10^8 years (with $t_{const} < 10^3$ yr).

Once the exploration and/or colonization of the Galaxy has begun, it can be modelled quite closely by the mathematical theory of island colonization – a theory developed fairly extensively by MacArthur & Wilson *(47,48)* – since the islands in the ocean are closely analogous to stars in the heaven, and the von Neumann machines are even more closely analogous to biological species. There are several general conclusions applicable to interstellar exploration and/or colonization which follow from this theory. First, there are two basic behavioral strategies, the *r*-strategy and the *K*-strategy, which would be adopted in different phases of the colonization. Here *r* is the net reproductive rate, and *K* is the carrying capacity of the environment. The *r*-strategy is one of rapid reproduction at the expense of all else, and it would be followed in the early stages of the colonization, while the *K*-strategy has a much smaller *r* and emphasis on securing the ecological niche in the target stellar system. The *K*-strategy would be adopted after the solar system had been colonized for some time, and would result in fewer probes being sent to other stars. In the past few centuries Western society has been an *r*-strategist, but as the carrying capacity of the environment is approached, it is beginning to adopt a *K*-strategy. Second, the MacArthur–Wilson theory suggests (49) that the fraction of probes reaching a distance *d* from the system of launch is $\sqrt{2/\pi}\,[\exp{(-d^2/2)}]/d$, which could result in an exploration rate of up to twice the value of 2.5×10^{-2} ly/yr with $v_{es} = 5 \times 10^{-2}\,c$ probes.

3   **Astrophysical constraints on the evolution of intelligent species**
     The probability that intelligent life which eventually attempts interstellar communication will evolve in a star system is usually expressed by the Drake equation:

$$p = f_p n_e f_l f_i f_c$$

where f_p is the probability that a given star system will have planets, n_e is the number of habitable planets in a solar system that has planets, f_l is the probability that life evolves on a habitable planet, f_i is the probability that intelligent life evolves on a planet with life, and f_c is the probability that an intelligent species will attempt interstellar communication within 5 billion years after the formation of the planet on which it evolved. The time limit in f_c is tacit in most discussions of extraterrestrial intelligence. Some time period of approximately 5 billion years must be assumed in order to use the Drake equation to estimate the number of existing civilizations. If, for example, f_c were a Gaussian distribution with maximum at $t = 30$ billion years and $\sigma = 1$ billion years, then we would be the only civilization in the galaxy. The probability estimates made below will hold if it is assumed that f_c is either sharply peaked at 5 billion years after planetary formation or a Gaussian distribution with $t_{peak} < 6$ billion years and $\sigma > 1$ billion years.

     The problem with the Drake equation is that only f_p – and to a lesser degree n_e – is subject to experimental determination. In order to measure a probability with a high degree of confidence, one must have a fairly large sample; for f_l, f_i, and f_c we have only one obvious case, the Earth (50). However, if one accepts the argument of the previous sections that any intelligent species which attempts interstellar communication will begin the galactic exploration program outlined within 100 years after developing the technology for interstellar communication, then the sample size is enlarged to include all those stellar systems older than $t_{age} = 5$ billion years + t_{ex}, where $t_{ex} \leq 300$ million years is the time needed to expand throughout the Galaxy. That is, the Drake probability p is less than or equal to $1/N$, where N is the number of stellar systems older than t_{age}, because all of these stars were, under the assumptions underlying the Drake equation, potential candidates to evolve communicating intelligent species, yet they failed to do so – had such species evolved on planets surrounding these stars within 5 billion years after star formation, their probes would already be present in the solar system, and these probes are not here *(14,15)*. Since f_p and n_e can in principle be determined by direct astrophysical measurement, the fact that extraterrestrial intelligent beings are not present in our solar system permits us to obtain a direct astrophysical measurement of an upper bound to the product $f_l f_i f_c$, which depends only on biological and sociological factors.

This argument assumes that the five probabilities of the Drake equation do not vary rapidly with galactic age. The available astrophysical evidence and most theories of the formation of solar systems indicate that this assumption is valid. The formation of solar systems requires that the interstellar gas be sufficiently enriched by 'metals' (those elements heavier than helium). Most experts *(29,51–53)* agree that a substantial fraction of existing metals were formed in massive stars very early in galactic history – during the first 100 million years of the Galaxy's existence – and the metal abundance has changed by at most a factor of about two since then. The evidence *(54,55)* gives a galactic age of between 11 and 18 billion years, and it is generally assumed *(52)* that the rate of star formation has been decreasing exponentially ever since the initial burst of heavy element formation. Existing stellar formation theory is unable to decide definitely if the so-called initial mass function – the number of stars formed per unit time with masses between m and $m + \Delta m$ – changes with time after the initial burst of massive stars *(51)*. Furthermore, it is not clear to what extent the earthlike planet formation rate depends on the metal abundance *(56,57)*. However, the observational evidence *(51)* (such as it is) does not indicate a large variation of the initial mass function of the earthlike planet formation rate with time. I shall thus assume that these are roughly constant, and most discussions of extraterrestrial intelligence make the same assumption *(58,59)*. The factors f_l, f_i, f_c should not depend strongly on the evolution of the Galaxy as a whole (see however *(60–62)*), and so can be regarded as constants. Since the Galaxy is between 11 and 18 billion years old, the number N of stars older than

5.3 billion years is about twice the number of stars formed after the Sun, and thus is approximately equal to the number of stars in the Galaxy, 10^{11}. Thus $p \leqslant 10^{-11}$. If we accept the usual values of $f_p = 0.1$ to 1 and $n_e = 1$ found in most discussions of interstellar communication (2,18), then $f_i f_i f_c \leqslant 10^{-10}$. The number of communicating civilizations now existing in our Galaxy is less than or equal to $p \times$ (number of stars in galaxy) = 1; that it to say, us.

This conclusion that we are the only technical civilization now existing in the Galaxy does not depend on any biological or sociological arguments except for the assumption that a communicating species would eventually begin interstellar travel; nor does it depend on f_p or n_e. It follows from just the interstellar travel assumptions, the assumption that the galactic environment has not changed by more than a factor of five during the history of the Galaxy, and the fact (?) that extraterrestrial probes are not present in our solar system.

4   Motivations for interstellar communication and exploration

It is difficult to construct a plausible scenario whereby an intelligent species develops and retains for centuries an interest in interstellar communications together with the technology to engage in it, and yet does not attempt interstellar travel. Even if we adopt the pessimistic point of view that all intelligent species cease communication efforts before developing von Neumann machines, either because of a loss of interest or because they blow themselves to bits in a nuclear war, the conclusion that we are the only intelligent species in the Galaxy with interest in interstellar communications is not changed. For in this case, the longevity L of a communicating civlization is less than or equal to 100 years (using our computer experts' opinions for the time needed to develop von Neumann machines), and since the Drake equation gives $n = R_* p L$ for the number of communicating civilizations in the Galaxy, we obtain $n = 10$, even if we use Sagan's optimistic estimate (2) of $R_* p = 1/10$. (The number R_* is the average rate of star formation.) This value of n is essentially the same as $n \sim 1$ obtained in the previous section, and in any case such short-lived civilizations would on the average be too far apart and exist for too short a time to engage in interstellar communication. If $L \geqslant 100$ years so that the species has time to develop probe technology, the value of L is irrelevant to the calculation of the number p. Once the probes have been launched, they will explore the Galaxy automatically; the death of the civilization that launched them would not stop them. We are thus left with the possibility that for some reason, intelligent beings with the technology and desire for radio communication do not use the exploration strategy because they *choose* not to do so, not because they are incapable of developing the technology.

There is no good reason for believing this. Virtually any reason for engaging in interstellar radio communication provides an even stronger argument for the exploration of the Galaxy. For example, if the motivation for communication is to exchange information with another intelligent species, then as Bracewell (*63,64*) has pointed out, contact via space probe has several advantages over radio waves. One does not have to guess the frequency used by the other species, for instance. In fact, if the probe has a von Neumann machine payload, then the machine could construct an artifact in the solar system of the species to be contacted, an artifact so noticeable that it could not possibly be overlooked. If nothing else, the machine could construct a 'Drink Coca-Cola' sign a thousand miles across and put it in orbit around the planet of the other species. Once the existence of the probe has been noted by the species to be contacted, information exchange can begin in a variety of ways. Using a von Neumann machine as a payload obviates the main objection (*65*) to interstellar probes as a method of contact, namely the expense of putting a probe around each of an enormous number of stars. One need only construct a few probes, enough to make sure that at least one will succeed in making copies of itself in another solar system. Probes will then be sent to the other stars of the galaxy automatically, with no further expense to the original species.

Morrison has expressed the opinion (*4*): '. . . once there is really interstellar communication, it may be followed by a ceremonial interstellar voyage of some special kind, which will not be taken for the sake of the information gained, or the chances for trade . . . , but simply to be able to do it, for one special case, where there is a known destination. That's possible, one can imagine it being done – but it is very unlikely as a search procedure'. However, if it is granted that a *single* probe is launched, for *any* reason, then with a von Neumann machine payload, the same probe can be used to start the galactic expansion program outlined in Section 2. While *en route* to a solar system known to be inhabited, the probe could make a stop-over at a stellar system along the way, make several copies of itself, refuel and then proceed on its way (or send one of the copies to the inhabited system). If the inhabited system is farther than 100 light years, and if $v_{es} \leq 0.1\,c$ and $f_{const} \leq 100$ years, then the time needed to reach the inhabited system is increased by less than 10 per cent, and one obtains the exploration and/or colonization of the entire Galaxy as a free bonus. Furthermore, because the inhabited system is so far away, *any* probe sent would have to be autonomous, which would mean a computer with a human-level intelligence, and capable of self-repair – which means that it would essentially be a von Neumann machine. Since its instrumentation makes *any* interstellar probe capable of galactic exploration, why not use it for that?

Consider the search strategy adopted by the *first* species interested in interstellar communication to arise in our Galaxy. Most likely it would be thousands or even millions of years before another such species arose. Even if another species arose simultaneously, the probability is only about 10^{-6} that it would be within 100 light years of the other species. Therefore, when the first species begins to signal, it will probably get no answer for thousands or millions of years. During this time it will be receiving no information on other stellar systems for its investment. If there remains strong interest in interstellar communication during this period, why should it not also launch a few probes? *Some* information on other systems would be guaranteed in 100 to 10^4 years, even if other intelligent beings are not discovered. Also, if there are other intelligent beings in the Galaxy, the von Neumann probes will eventually find them, even if they are intelligent beings who would never develop on their own an interest in interstellar communication. With radio waves and a null result, there is always the possibility that the wrong frequency has been chosen, that some other means than radio waves has been used by the other species, etc. There is no such problem with probes.

If human history is any guide, this first species will launch a probe rather than make radio beacons in the first place. In the early part of this century, when Lowell had convinced many that there were intelligent beings on Mars, but when interplanetary rocket probes were regarded as a ridiculous fantasy, the Harvard astronomer W. H. Pickering pointed out (*66*) that communication with these beings was possible with a mirror one-half square mile in area: '[it] would be dazzlingly conspicuous to Martian observers, if they were intellectually and physically our equals'. If we were content to use such a device to learn about Mars from these hypothetical Martians, we would still know little about Mars. Instead, we sent probes, and Sagan's recent proposals (*67*) for advanced Mars probes are robots with manipulative ability and a considerable degree of artificial intelligence – they are a step in the direction of a von Neumann machine.

If we assume that a behaviour pattern which is typical not only of *Homo sapiens* but also of all other living things on our planet would also be adopted by any intelligent species (to deny this would be to deny the assumption of mediocrity), then we would conclude that a sufficiently advanced intelligent species would launch von Neumann probes. All living things have a dispersal phase (*68*), in which they tend to expand into new environments, for the dispersal behaviour pattern is obviously selected by natural selection. The expansion is generally carried out to the limit imposed by their genetic constitution. In intelligent species, this limit would be imposed by the level of technology (*69,70*), and we would expect the dispersal behaviour pattern to be present in at least some groups of an intelligent species. We should therefore expect that at least some groups of the species would attempt an expansion into the Galaxy, and the construction of only one successful von Neumann probe would be sufficient for this. By launching such a probe and using it to colonize the stars, a species increases the probability that it will survive the death of its star, nuclear war, etc. Note that it need not take territory away from another species (intelligent or not) to accomplish this purpose. The species could, for example, restrict itself to the construction of O'Neill colonies around stars with no living things on their planets.

It is possible that an intelligent species which develops a level of technology capable of interstellar communication would decide not to build von Neumann machines because they would be afraid that they would lose control of the machines. Since no reproduction can be perfect, it is possible that the program which keeps the von Neumann probes under the control of the intelligent species could accidentally be omitted during the reproduction process, with the result that the copy goes into business for itself. This problem can be avoided in three ways. First, the program which keeps the probe under control can be so integrated with the total program that its omission would cause the probe to fail to work at all. This is analogous to the constraints imposed on the cells used in recombinant DNA technology. Second, the intelligent species could program the probes to form colonies of the intelligent species in the stellar system reached by the probes. These colonies would be able to destroy any probes which slipped out of control. Third, the intelligent species might not care if the von Neumann machines slipped out of control. After all, a von Neumann machine would be an intelligent being in its own right, an intelligent being made of metal rather than flesh and blood. The rise of human civilization has been marked by a decline in racism, and an extension of human rights – which include freedom – to a wider and wider class of people. If this trend continues and occurs in the cultures of all civilized beings, it seems likely that von Neumann machines would be recognized as fellow intelligent beings, beings which are the heirs to the civilization of the naturally evolved species that invented them, and with the right to the freedom possessed by the inventing species. If on the other hand, the intelligent species retained their racism, it seems likely that they would regard other 'flesh and blood' intelligent species as 'non-people'. If so, then they would either wish to avoid communication altogether (lest it 'pollute' their culture with alien ideas), or else launch von Neumann machines to either colonize the Galaxy for themselves (lest it be done by 'non-people' who would crowd them out) or to destroy these other intelligent species. For example, this colonization or destruction would be their best strategy if they believed that the biological 'exclusion principle', which says (71,72) two species cannot occupy the same ecological niche in the same territory, applies to intelligent species. With the advent of the O'Neill colony, the ecological niche occupied by an intelligent species would consist of the entire material resources of a solar system. The ecological niches of two intelligent species would have to overlap. In any case, the von Neumann probes would be launched. If a species was not afraid of alien ideas itself, but was reluctant to contaminate the culture of another species with its own culture, then it should not attempt radio contact. However, with probes it would be possible to study an alien species without it becoming aware of the species which was studying it.

A final possibility to be considered is what I have hitherto denied, namely that perhaps the von Neumann probes of an extraterrestrial intelligent species *are* present in our solar system. If a probe had just arrived, there would as yet be no evidence for its presence. The probability that a probe arrived for the first time within the past 20 years is 10^{-9} [= 20/(age of Galaxy)]. Thus the probability that extraterrestrial intelligent beings exist but their probes have just arrived is actually greater than the calculated probability $f_i f_i f_c$ that they evolve. Another possibility would be that they are here but have decided for some reason not to make their presence known; this is the so-called zoo hypothesis (73). Kuiper & Morris (13) have proposed testing this hypothesis by attempting to intercept radio communications between beings in our solar system and the parent stars. Another possible test would be to search for the construction activities of a von Neumann machine in our solar system. For example, one could look for the waste heat from such activities. As Dyson has pointed out (11,74), this heat would give rise to an infrared excess, and the most likely place to look for a von Neumann probe would be the asteroid belt where material is most readily available. (It is amusing that infrared radiation of astronomical origin does come from the asteroid belt (75).) If such a von Neumann probe were present in the solar system and if a large number of mutually intercommunicating intelligent species existed who were interested in studying us, we would expect the von Neumann machine to construct members of each of these species, together with spaceships, one appropriate type for each of the species.

We would thus expect to see a wide variety of species and spaceships on Earth studying us (*76*). But no extraterrestrial ships of any type are seen (*14,15*). Furthermore, if intelligent beings existed, it is likely that their probes would have arrived a billion years ago when there was nothing on Earth but one-celled organisms, and hence they would have no reason to hide their technology. The entire asteroid belt would be artifacts by now. Thus the evidence is enormous that extraterrestrial intelligent beings do not exist.

But the evidence is not utterly conclusive; beings with extremely advanced technology could be present in our solar system and make their presence undetectable should they wish to do so. The point is that a belief in the existence of extraterrestrial intelligent beings anywhere in the galaxy is not significantly different from the widespread belief that UFOs are extraterrestrial spaceships. In fact, I strongly suspect the psychological motivation of both beliefs to be the same, namely 'The expectation that we are going to be saved from ourselves by some miraculous interstellar intervention . . .' (quoted from (*77*), page 272).

As discussed in ref. (1), belief in extraterrestrial intelligent beings is associated with a belief in the immensity of the Cosmos: if there is a huge number of habitable planets, is it plausible that there is only one inhabited planet? I would contend the answer is yes. Wheeler has argued (*78*) that if the Universe were much smaller than it is, it would terminate in a final singularity before intelligent life would have time to evolve. This is an example of an 'Anthropic Principle' argument. The Anthropic Principle (*79–81*) states that many aspects of the Universe are determined by the requirement that intelligent life exists in it. Thus the Universe must contain 10^{20} stars in order to contain a single intelligent species. We should not therefore be surprised if indeed it contains only one.

References

(1) The singularity *v.* the plurality of inhabited worlds is a thema–antithema pair in sense of Holton (*The Scientific Imagination: Case Studies*, Cambridge University Press, 1978) and so has a long history in science. The medieval church contended that man was unique, but from the Renaissance (e.g. Bruno) to the middle of the nineteenth century, the notion of the plurality of inhabited worlds held sway. A belief in the plurality of inhabited worlds is generally associated with three other beliefs: one, a belief in spontaneous generation; two, a belief in a large number of worlds; and three, a belief in a 'Chain of Being' which asserts intelligence must exist if low-level organisms do. [See Lovejoy, A. O., 1936. *The Great Chain of Being*, Harvard University Press, for a discussion of belief three.] The best known works advocating plurality were Fontenelle's *Dialogues on the Plurality of Worlds* (1686); Huygens' *Theory of the Universe* (1690); and Brewster's *More Worlds*

148   *Frank J. Tipler*

Than One (1854). A critique of these views was given by Whewell in his *The Plurality of Worlds* (1855), but a decline of the belief in inhabited worlds began with the rise of the theory of evolution (see in particular evolutionist Alfred R. Wallace's critique *Man's Place in Nature* (1904) and also *Fortnightly Review*, **73** (n.s.), 395 (1903); **74**, 380 (1903)), the decline of belief in Laplace's nebular hypothesis, and Pasteur's disproof of spontaneous generation. The modern belief in the plurality of worlds comes from two sources: Lowell's arguments for intelligent life on Mars (see Hoyt, W. G., 1976. *Lowell and Mars*, University of Arizona Press, Tucson), and the classic paper by Cocconi and Morrison (written after the revival of the nebular hypothesis and the 'spontaneous generation' experiments of Stanley Miller) pointing out that we could detect extraterrestrial civilizations by means of microwaves. (The possibility of communication with extra-solar civilizations by means of radio had actually been discussed much earlier by Barnes, E. W., 1931. *Nature*, **128**, 722.)

(2) Shklovskii, I. S. & Sagan, C., 1966. *Intelligent Life in the Universe*, Dell, New York.

(3) Drake, F. D., 1960. *Intelligent Life in Space*, Macmillan, New York.

(4) Morrison, P., 1974. In: *Interstellar Communication: Scientific Perspectives*, ed. Ponnamperuma, C. & Cameron, A. G. W., Houghton-Mifflin, Boston.

(5) Dobzhansky, T., 1972. In: *Perspectives in Biology and Medicine*, **15**, 157. Dobzhansky, T., 1973. *Genetic Diversity and Human Equality*, pp. 99–101, Basic Books, New York.

(6) Simpson, G. G., 1964. *This View of Life*, chapters 12 and 13, Harcourt, New York.

(7) Francois, J., 1977. *Science*, **196**, 1161; see also Mathew, W. D., 1921. *Science*, **54**, 239.

(8) Dobzhansky, T., Ayala, F. J., Stebbins, G. L. & Valentine, J. W., 1977. *Evolution*, Freeman, San Francisco.

(9) Mayr, E., 1978. *Scientific American*, **239**, 46 (September).

(10) Fermi, E.: quoted on page 495 of: Sagan, C., 1963. *Planetary Space Science*, **11**, 485.

(11) Dyson, F. J., 1966. In: *Perspectives in Modern Physics: Essays in Honor of Hans A. Bethe*, ed. Marshak, R. E., Wiley, New York.

(12) Hart, M. H., 1975. *Quarterly Journal of the Royal Astronomical Society*, **16**, 128.

(13) Kuiper, T. B. H. & Morris, M., 1977. *Science*, **196**, 616.

(14) Klass, P. J., 1974. *UFOs Explained*, Random House, New York.

(15) Menzel, D. H. & Taves, E. H., 1977. *The UFO Enigma*, Doubleday, Garden City, New York.

(16) Shklovskii, I. S. & Sagan, C., *op. cit.*, *2*, chapter 25.

(17) Sagan, C., 1977. *The Dragons of Eden*, p. 239, Ballantine, New York.

(18) *Project Cyclops* (Report CR114445, NASA Ames Research Center, Moffett Field, California, 1971).

(19) *The Search for Extraterrestrial Intelligence: SETI* (NASA report SP-419, 1977).

(20) Michie, D., 1973. *Nature*, **241**, 507.

(21) Firschein, O., Fischler, M. A. & Coles, L. S., 1973. In: *Third International Joint Conference on Artificial Intelligence*, Stanford University. This reference actually gives the opinions of leading computer scientists as to when computers with human-level intelligence and manipulative ability will be developed. This technology seems to me to be roughly comparable to von Neumann machine technology, so I use this number as my estimate for how long it will be before we develop von Neumann machines. No explict mention was made of von Neumann machines in references *20*, *21*, or *22*.

(22) Minsky, M., 1973. In: *Communication with Extraterrestrial Intelligence*, p. 160, ed. Sagan, C., MIT Press, Cambridge.

(23) Neumann, J. Von, 1966. *Theory of Self-Reproducing Automata*, edited and completed by Burks, A. W., University of Illinois Press, Urbana.

(24) Arbib, M. A., 1969. *Theories of Abstract Automata*, Prentice-Hall, Englewood Cliffs, N.J.; see also the Arbib article in Ponnamperuma & Cameron, *op. cit.*, *4*.

(25) Stuhlinger, E., 1964. *Ion Propulsion for Space Flight*, McGraw-Hill, New York.

(26) Wright, J. L. & Warmke, J. M., 1976. *Solar Sail Mission Applications*, JPL preprint 76-808, AIAA/AAS 1976 San Diego Astrodynamics Conference.

(27) Chapman, C. R., 1975. *Scientific American*, **232**, 24 (January); Skinner, B. J., 1976. *American Scientist*, **64**, 258; Hughes, D. W., 1977. *Nature*, **270**, 558.

(28) Abt, H. A., 1977. *Scientific American*, **236**, 96 (April); Batten, A. H., 1973. *Binary and Multiple Systems of Stars*, Pergamon Press, New York; Dole, S. A., 1964. *Habitable Planets for Man*, Blaisdell, New York.

(29) Truran, J. W. & Cameron, A. G. W., 1971. *Astrophysical Space Science*, **14**, 179; Cameron, A. G. W., *op. cit.*, *4*.

(30) O'Neill, G. K., 1974. *Physics Today*, **27**, 32 (September); 1975. *Science*, **190**, 943; 1977. *The High Frontier*, Morrow, New York.

(31) Price, C. C., 1965. *Chemical Engineering News*, **43**, 90 (September 27); Price, C. C. (ed.), 1974. *Synthesis of Life*, pp. 284–286, Dowden, Hutchinson & Ross, Stoudsburg, PA.

(32) Danielli, J. F., 1972. *Bulletin of Atomic Scientists* (December), pp. 20–24 (also in Price, C. C., *op. cit.*, *31*); Jeon, K. W., Lorch, I. J. & Danielli, J. F., 1970. *Science*, **167**, 1626.

(33) Grobstein, C., 1979. *Scientific American*, **240**, 57 (June).

(34) Eiseley, L., 1970. *The Invisible Pyramid*, pp. 78–80, Scribner's, New York.

(35) Crick, F. H. C. & Orgel, L. E., 1973. *Icarus*, **19**, 341.

(36) Sneath, P. H. A., 1962. *Nature*, **195**, 643.

(37) Seibert, M., 1976. *Science*, **191**, 1178; 1977. *In Vitro*, **13**, 194.

(38) Cravalho, E. G., 1975. *Technology Review*, **78**, 30 (October).

(39) Parkes, A. S., 1965. *Sex, Science and Society*, Oriel Press, London.

(40) Hunter, II, M. W., 1967. *AAS Science and Technology Series*, **17**, 541.

(41) Morgenthaler, G. W., 1969. *Annals of the New York Academy of Sciences*, **163**, 559.

(42) Helton, M. R., 1977. *Jet Propulsion Laboratory Interoffice Memorandum*, 312/774-173 (June 21).

(43) Bond, A. *et al.*, 1978. *Project Daedalus* (Special suppl. *Journal of the British Interplanetary Society*).

(44) *Op. cit.*, *19*, p. 108.

(45) Dyson, F. J., 1969. *Annals of the New York Academy of Sciences*, **163**, 347; see also Spencer, D. F. & Jaffe, L. D., 1962. *Jet Propulsion Laboratory*, Preprint.

(46) Dyson, F. J., 1974. Quoted in Berry, A., 1974. *The Next Ten Thousand Years*, p. 125, New American Library, New York.

(47) MacArthur, R. H. & Wilson, E. O., 1967. *The Theory of Island Biogeography*, Princeton University Press, Princeton.

(48) Wilson, E. O., 1975. *Sociobiology*, Harvard University Press, Cambridge.

(49) Wilson, E. O., *op. cit.*, *48*, p. 105.

(50) Sagan, C., (ed.) 1973. *Communication with Extraterrestrial Intelligence*, MIT Press, Cambridge.

(51) Trimble, V., 1975. *Reviews of Modern Physics*, **47**, 877.

(52) Audouze, J. & Tinsley, B. M., 1976. *Annual Reviews of Astronomy and Astrophysiology*, **14**, 43.

(53) Penzias, A. A., 1978. *Comments on Astrophysiology*, 8, 19.

(54) Browne, J. C. & Berman, B. L., 1976. *Nature*, 262, 197.

(55) Hainebach, K. L. & Schramm, D. N., 1976. *Enrico Fermi Institute Preprint*, No. 76–16, University of Chicago Press.

(56) Talbot, R. J., 1974. *Astrophysical Journal*, 189, 209; Talbot, R. J. & Arnett, W. D., 1973. *Astrophysical Journal*, 186, 51.

(57) Barry, D. C., 1977. *Nature*, 268, 509.

(58) *Op cit.*, 18, p. 25.

(59) Kreifeldt, J. G., 1971. *Icarus*, 14, 419.

(60) *Op. cit.*, 2, p. 100.

(61) Alvarez, W., Alvarez, L. W., Asaro, F. & Michel, H. V., 1979. Experimental Evidence in Support of an Extraterrestrial Trigger for the Cretaceous-Tertiary Extinctions, *American Geophysical Union Transcripts*, 60, 734.

(62) Verschuur, G. L., 1977. Preprint: *Will We Ever Communicate With Extraterrestrial Intelligence?*, University of Colorado.

(63) Bracewell, R. N., 1960. *Nature*, 186, 670; reprinted in: Cameron, A. G. W. (ed.), 1963. *The Search for Extraterrestrial Life*, Benjamin, New York.

(64) Bracewell, R. N., 1975. *The Galactic Club*, Freeman, San Francisco.

(65) *Op. cit.*, 19, p. 108.

(66) Pickering, W. H., 1909. *Popular Astronomy*, 17, 495; reprinted in: Pickering, W. H., 1921. *Mars*, Gorham Press, Boston.

(67) Sagan, C., 1977. Quoted in: *Technology Review*, 79, 14 (May).

(68) Dobzhansky, T., 1970. *Genetics of the Evolutionary Process*, p. 278, Columbia University Press, New York.

(69) Morison, S. E., 1965. *Portuguese Voyages to America in the Fifteenth Century*, pp. 11–15, Octogon, New York.

(70) Davies, K., 1974. *Scientific American*, 231, 92 (September).

(71) Mayr, E., 1970. *Populations, Species, and Evolution*, p. 48, Harvard University Press, Cambridge.

(72) May, R. M., 1978. *Scientific American*, 239, 160 (September).

(73) Ball, J. A., 1973. *Icarus*, 19, 347.

(74) Dyson, F. J., 1960. *Science*, 131, 1667.

(75) Low, F. J. & Johnson, H. J., 1964. *Astrophysical Journal*, 139, 1130.

(76) Thus the argument by Chiu, H. Y., 1970. *Icarus*, 11, 447, that UFOs could not be explained as spaceships, because the observed visit rate would require too much material in the form of spaceships, is incorrect. Only one von Neumann probe need be sent to each solar system, and the material used for constructing the spaceships in each solar system could be re-used.

(77) Sagan, C., 1972. In: *UFOs – A Scientific Debate*, ed. Sagan, C. & Page, T., Norton, New York; this hope of extraterrestrial salvation is particularly obvious in Drake, F., 1976. *Technology Review*, 78, 22 (June).

(78) Wheeler, J. A., 1977. In: *Foundational Problems in the Special Sciences*, eds. Butts, R. E. & Hintikka, J., Reidel, Dordrecht.

(79) Carter, B., 1974. In: *Confrontation of Cosmological Theories with Observational Evidence*, ed. Longair, M. S., Reidel, Dordrecht.

(80) Carr, B. J. & Rees, M. J., 1979. The anthropic principle and the structure of the physical world, *Nature*, 278, 605–12.

(81) Barrow, J. D. & Tipler, F. J., 1985. *The Anthropic Cosmological Principle*, Oxford University Press, in press.

The solipsist approach to extraterrestrial intelligence

CARL SAGAN
WILLIAM I. NEWMAN

One of the distinctions and triumphs of the advance of science has been the deprovincialization of our world view. In the sixteenth century there were battles over whether the Earth is at the centre of the Solar System; in the seventeenth century about whether the stars are other suns; in the nineteenth century, about whether the Earth is much older than real or mythical human history; in the eighteenth to twentieth centuries about whether the spiral nebulae are other galaxies something like the Milky Way, and about whether the Sun is at the centre of the Milky Way; and in the nineteenth and twentieth centuries about whether human beings have arisen and evolved as an integral part of the biological world, and whether there are priviledged dynamical frames of reference. These deep questions have generated some of the major scientific advances since the Renaissance. Every one of them has been settled decisively in favour of the proposition that there is nothing special about us: we are not at the centre of the Solar System; our planet is one of many; it is vastly older than the human species; the Sun is just another star, obscurely located, one among some 400 billion others in the Milky Way, which in turn is one galaxy among perhaps hundreds of billions. We humans have emerged from a common evolutionary process with all the other plants and animals on Earth. We do not possess any uniquely valid locale, epoch, velocity, acceleration, or means of measuring space and time.

The latest issue in this long series of controversies on our place in the Universe properly concerns the existence of extraterrestrial intelligence. Despite the utter mediocrity of our position in space and time, it is occasionally asserted, with no sense of irony, that our intelligence and technology are unparalleled in the history of the cosmos. It seems to us more likely that this is merely the latest in the long series of anthropocentric and self-congratulatory pronouncements on scientific issues that dates back to well before the time of Claudius Ptolemy. The history of deprovincialization of course does not *demonstrate* that there are intelligent beings elsewhere. But, at the very least, it urges great caution in accepting the arguments of those who assert that no extraterrestrial intelligence exists. The only valid approach to this question is experimental.

The most elaborate recent exposition of the solipsist world view is that of Tipler (1980; 1981a,b,c,d; 1982a,b). Seeking, in effect, a universal principle to explain the apparent absence of extraterrestrial beings on Earth, he contends that if extraterrestrial intelligent beings exist, their manifestations will be obvious; conversely, since there is no evidence of their presence, they do not exist. But absence of evidence is not evidence of absence. Because some experts in artificial intelligence propose that general purpose self-replicating robots will be devised on Earth within a century, Tipler concludes that any extraterrestrial civilization, only a little more advanced than we, will necessarily infest the Galaxy in a few million years with exponentiating devices. Adopting a conservative assumption on velocity of 3×10^{-4} light yr yr^{-1}, Tipler deduces a replication time for his 'von Neumann machines' $\sim 10^4$ yr. In several places in his discussion he imagines that a von Neumann machine, landing on a virgin world, makes no more than a few copies of itself. The reason for such reproductive restraint is never mentioned. With the implacable dedication to self-replication with which these machines are supposedly endowed, thousands or millions of replications per world seems much more likely. However, with any plausible initial mass for such a device, and with even one copy per reproductive event, the entire mass of the Galaxy would be converted into von Neumann machines within a few million years of their invention. (Stars are, after all, excellent sources of the ^{3}He that Tipler imagines the von Neumann machines to hunger for.) For example, with a modest mass $\sim 10^9$ g per von Neumann machine they will consume $10^9 \, 2^r$ g in r generations. If $r \sim 150$, this is $\sim 10^{54}$ g $\geqslant$ the mass of any known galaxy. The conversion will take $<1.5 \times 10^6$ yr. These considerations apply equally to other galaxies, none of which appears to have been converted into von Neumann machines. Thus we can derive a still stronger consequence from these dubious arguments than Tipler does: if there is anywhere an intelligence more advanced than ours, then substantial volumes of the Universe will have been fundamentally reworked, in apparent contradiction to observation. It follows that ours is the most advanced civilization not merely in our Galaxy, but in the cosmos; of all the $\sim 10^{23}$ planets which are implicit in Tipler's arguments, and in the ~ 15 billion year history of the Universe, ours is the only world on which an advanced technology has evolved. This naturally raises a question. Considering the history of solipsist argument, which is more likely: that in a 15 billion-year-old contest with 10^{23} entrants, we happen, by accident, to be first or that there is some flaw in Tipler's argument?

Let us for the moment set aside any doubts about whether von Neumann machines of the sort Tipler describes are technologically feasible or will have their development supported by any advanced society. Either there are entirely reliable restraints on their rates and sites of replication, or there are not. In the former case, our Solar System may have been intentionally bypassed because of the evolution of intelligence here, or for many other possible reasons, and the problem Tipler poses vanishes. In the latter case, these implacable replicators will not stop until the entire Universe has been converted into $\sim 10^{47}$ von Neumann machines, which then presumably cannibalize each other. If anything like this were a real danger, an emerging interstellar civilization would be wise, as a matter of self-preservation, to take steps to prevent it. No civilization could be sure of the ultimate fidelity over $>10^{6}$ yr of the self-replication programs of the von Neumann machines, and whether they ultimately might pose as grave a threat to the planets of their builders as to any other world. Tipler's assurance that the problem is comparable to, and as easily remedied as, that posed by recombinant DNA is unconvincing. In our problem there is no host specificity: the von Neumann machines are imagined able to enter any random planetary system and set about reproducing themselves in less than the transit time. Thus, the prudent policy of any technical civilization must be, with very high reliability, to prevent the construction of interstellar von Neumann machines and to circumscribe severely their domestic use. If we accept Tipler's arguments, the entire Universe is endangered by such an invention; controlling and destroying interstellar von Neumann machines is then something to which every civilization – especially the most advanced – would be likely to devote some attention. If the first galactic civilizations were not so careless as to have overlooked the cosmic danger that Tipler has called to our attention, it seems that we can safely dismiss the spectre of exponentiating machines that eat stars and worlds, and consider what other principles with some claim to universality might explain the apparent absence of extraterrestrials on Earth.

Thus, we return to the more 'conventional' scenarios of biological and mechanical beings, replicating together, slowly expanding in an irregular propagation wavefront to explore and possibly colonize other worlds, a circumstance we have discussed in some detail elsewhere (Newman & Sagan 1981, henceforth, Paper I). In order to model galactic colonization, it is convenient to develop a colonization strategy that is applicable generally and in which differences in strategy are described through choices of numerical parameters. Once such a formulation is complete, mathematical techniques can be employed to transform this microscopic description into a macroscopic one describing colonization on a galactic scale. The search strategy we employ in Paper I is characterized by three assumptions that we believe to be plausible:

(1) colonization ventures are more likely to be launched from worlds with substantial populations, rather than from fledgling, newly established colonies;

(2) colonies are more likely to be established on unpopulated or sparsely populated worlds than on well-developed, pre-existing colonies or planets with advanced indigenous organisms; and

(3) colonization drives will be mounted towards the nearest available colonization sites, not toward very remote worlds which would be rendered less attractive by economic constraints and by motivational questions.

Taken together with a population dynamic representation for *in situ* population growth and saturation, we constructed a mathematical description of this colonization strategy. Jones (1978) earlier developed a Monte Carlo computer code for such a colonization strategy. Our treatment is a rigorous analytic description of the problem, applicable to a very wide range of parameter selections, that reproduces Jones' results for his choice of parameters. Our preferred parameterization employs numerical values consistent with long-term human history, while, in our opinion, Jones' preferred values invoke atypically high values of the population growth and emigration rates. We discuss this point further below. On a microscopic level, our strategy corresponds to a Markov process. When the emigration time-scale is long relative to the e-folding time for population growth, the mathematical description must be retained in discrete form – as was the case for some of our models and for all of Jones'. But when the emigration time-scale is short relative to the population growth time, the description can be reduced to a non-linear partial differential equation – as was the case for another subset of our models. Both cases represent a directed form of Brownian motion on a microscopic scale, and to non-linear diffusion on a macroscopic scale. We refer to these two cases of our multiplicative diffusion model as the discrete and continuous limits, respectively. There are important differences that render the behaviour of our model very different from the usual attributes of a diffusive process.

Some readers of our paper (e.g. Tang 1982) have assumed, despite our lengthy elaboration of the differences between linear and non-linear diffusion models, that our results are prone to the anomalies inherent in the former. Since the model incorporates an increased likelihood of a colony being launched from a well-populated world to an unpopulated one, the propagation is directed away from population centers to virgin worlds (a mechanism called 'directed motion' by Gurney & Nisbet 1975 in application to comparable terrestrial biological problems). The directivity imparted by the model into the emergent population flow produces a *nonrandom dispersal*. Spatial and temporal variability in the habitability of worlds emerges in the form of strong fluctuations in the effective diffusion coefficient and provides for substantial *anisotropy* in the resulting distribution of colonies. Our model, therefore, describes all but the mathematically most catastrophic situations, and different colonization scenarios are reduced to the selection of numerical parameters.

Tipler claims that our mathematical development, although appropriate to animal populations, is inappropriate to humans. He guesses that the colonization of North America proceeded at a velocity $v \cong 6$ miles yr^{-1}. The corresponding step size, Δx, scales as

$$\Delta x = v(t_{\text{travel}} + t_{\text{growth}}) \tag{1}$$

(cf. equation 71, Paper I). Here t_{travel} is a measure of the travel time from a settlement to a new colonization site, a matter of only a few decades at $v \sim 0.1\ c$. On the other hand, t_{growth} is a measure of the time-scale for population increase due to the combined influence of natural population growth (i.e. the excess of births over deaths) and of immigration. (Since $t_{\text{growth}} \gg t_{\text{travel}}$, we can safely ignore the latter.) Much of our paper (including the appendix) dealt with the computational details of establishing t_{growth} from the associated population growth rate γ and the specific emigration rate ψ. In assessing probable values of γ and ψ for interstellar colonization, we examined the historical record. The values we chose corresponded to colonization as a means of exploration, as an outlet for national or religious zeal, as a tool for plunder, but not as a means of relieving the burdens posed by excess population in the source nation. The remarkable early growth in the population of the United States was the direct consequence of the overwhelming number of exiles who came to America seeking relief from an overcrowded, chaotic and hungry Europe following the Industrial Revolution. Our interstellar choices for γ and ψ do not describe and were not meant to describe this mass exodus. Our objective was to select values of γ and ψ appropriate to the age of exploration on Earth and, by inference, to the colonization of the stars.

Adopting numerical values for ψ and γ appropriate to the age of exploration prior to the demographic transition, and with his assumed value $v = 6$ miles yr^{-1}, Tipler obtains a time-scale ~ 1000 yr, and $\Delta \times \sim 10\,000$ miles, which he describes, correctly, as 'ridiculous'. But this value of t_{growth} is inappropriate to post-1750 America (as we have seen) and the value of v chosen, although quite reasonable after the American Revolution, is a tremendous overestimate for the United States prior to 1750. If $v \sim 6$ miles yr^{-1} were appropriate during the preceding 250 years, a coastal strip of depth $250 \times 6 = 1500$ miles would have been inhabited at or near the population saturation density. In contrast, the original 13 colonies, occupying a very narrow coastal strip, had a population under two million in 1750 (McEvedy & Jones 1978). Tipler's attempt to demonstrate a *reductio ad absurdum* in our non-linear diffusion model fails.

A more nearly correct determination of Δx can be accomplished by the use of appropriate parameters. We consider post-1750 America and compare it with Ireland around 1850. With the dramatic decline in the death rate precipitated by advances in public health and nutrition, the population growth rate in Ireland had risen to 0.01 yr^{-1} (Davis 1974). The Irish potato blight of the 1840s caused widespread famine and produced a flood of immigrants to the US. Indeed, ψ actually exceeded γ, and the overall population of Ireland experienced a continuous decline (although it was the only European country to do so). The natural population growth rate γ of eighteenth century America was 0.03 yr^{-1} (McEvedy & Jones 1978). As this unprecedented growth diminished in the early nineteenth century, the slack was taken up by the exodus from Ireland and elsewhere. As a consequence, the growth of America's population due to a combination of natural growth and immigration continued at the phenomenal rate of 0.03 yr^{-1} until this century, when it declined by a factor of 2. An aggregate population growth rate for the period 1700 to 1950 is slightly in excess of 0.02 yr^{-1}, implying a value for t_{growth} of ≤ 50 yr. The corresponding value of Δx, a measure of the width of the colonization front in our multiplicative diffusion model, is therefore ~ 6 miles yr^{-1}/0.02 yr^{-1} = 300 miles. It is not coincidental that this length scale is of the same order as the distance between major population centers in the continental United States.

Finally, Tipler indicates (without explanation) his dissatisfaction with our choice of what he calls t_{const} (the time-scale for a new colony to become sufficiently large to support a colonization venture of its own). This timescale, as explained in Paper I, scales as

$$t_{const} \sim (\ln \alpha)/\gamma \qquad (2)$$

where α is the ratio of the population at the time of launching a new colony of generation $n + 1$ to the population of the source colony of generation n. (Implicitly, this expression assumes exponential population growth at a rate γ.) Since the logarithmic factor is $\cong 10$, our disagreement focuses on the choice of γ. We prefer $\gamma \sim 6 \times 10^{-4}\,\mathrm{yr}^{-1}$, a value that describes the growth of the human population since the dawn of civilization to the middle of the eighteenth century; it corresponds to $t_{const} \sim 2 \times 10^4\,\mathrm{yr}$ (not $10^5\,\mathrm{yr}$ as Tipler claims). Tipler would prefer to use a substantially higher value for γ, perhaps one more representative of the recent global growth rate of $0.02\,\mathrm{yr}^{-1}$ (which demographers regard as aberrational). We have pointed out that at that level of fecundity, a hypothetical Adam and Eve could populate an Earth-like planet to its carrying capacity in a little over 1000 yr. Historical precedent argues strongly for much lower values of γ.

For our values of γ and ψ, it follows (equation 84, Paper I) that the time for a colonization wave to cross the Galaxy, allowing for discretization ($\bar{v} \sim 30$) in the solution to our equations, is 7.5×10^8 yr; or, with no allowance for discretization ($\bar{v} \sim 2$) $\sim 10^{10}$ yr. (We stress again that spacecraft velocities have little bearing on the solution to this problem.) The times would be several orders of magnitude longer if the colonizing societies adopted zero population growth. The galaxy-crossing time is proportional to $(\psi\gamma)^{-1/2}$, so our results are not very sensitive to the values of growth and emigration rates. It might be argued that the values of ψ chosen are improbably *large*. In this case the galaxy-crossing time will be longer still. But it is difficult to be certain about what progress in interstellar spaceflight will be exhibited by civilizations a million years, say, in our technological future. Even with the numbers selected, the galaxy-crossing times calculated with and without discretization correspond, respectively, to three galactic rotations and to the age of the Milky Way Galaxy itself. We believe that these are the most likely values of the time-scale required, in the absence of impediments from other civilizations, to populate every habitable world and to establish a full galactic empire.

For convenience, we refer to our value for the galaxy-crossing time as 'one billion years'. The establishment of galactic hegemony requires a perseverance to the task for a period of a billion years. The steadfastness of such a commitment seems remarkable. Given a doubling time of some technological figure of merit of, say, 30 yr – a rough index for the present epoch – the technology will advance in a thousand years by a factor of 10^9. With modest allowances for the nature of the technological advance in a thousand years – much less a billion – we can well imagine much more exciting and fulfilling objectives for an advanced civilization than strip-mining or colonizing every planet in sight. One of us proposed many years ago (Sagan 1973) that civilizations only a thousand years in our technological future become disinterested even in communicating with civilizations as backward as ours. A society thousands or millions of years more advanced than we are, beneficiaries of an exponentiating technology over all those intervening millennia, will not be engaged in a simple extrapolation of our activities, or be driven by our motives. It seems to us quite unlikely that an advanced technological civilization, undergoing continued biological and psychological as well as scientific development, will persevere in such imperialist designs for a billion years. Some hint of the possible concerns and enterprises of a very advanced society have recently been mentioned by Frautschi (1982). The idea that there will be no new and more compelling challenges for such civilizations than relentless galactic colonization represents a serious failure of the imagination – although it is a natural enough notion were we to extrapolate carelessly from recent human history.

But the colonization of our region of the Galaxy does not require that every advanced galactic civilization colonizes; only that one does. If there are abundant civilizations in the Galaxy, their absence on Earth requires a principle of universality so compelling as to admit essentially no exceptions. It is possible that the enormously greater challenges that are likely to be uncovered by an advanced technological civilization in less than the galaxy-crossing time provide a sufficient principle of universality. But there are a number of other candidate principles summarized in Paper I, one of which we wish to stress: the intrinsic instability of societies devoted to an aggressive galactic imperialism. (A culture that gave a wide berth to planetary systems in which life is evolving would, of course, pose no contradictions to the apparent absence of extraterrestrials in the Solar System; such a civilization could have occupied all the remaining planetary systems in our spiral arm and we would be none the wiser.) We need not speculate on the nature of future societies or extrapolate into the indefinite future in order to see why such societies might be self-limiting. We need only look around us. Since 1945 there has been a steep and monotonic increase in the number of nuclear weapons and the efficacy of their delivery systems. The present global arsenal of some 50 000 nuclear weapons of all types and more than 17 000 targeted strategic weapons (Sivard 1982) is more than adequate to destroy our technical civilization and perhaps (although this is not essential to our argument), through synergistic effects, to destroy our species as well. There are at present at least six nations possessing nuclear weapons and their delivery systems and three or four commonly mentioned additional candidate states.

Because of the widespread availability of weapons-grade fissionable materials and increasing world-wide competence in the relevant technology, the number of nations with nuclear weapons should increase sharply in the next few decades, if no new and major precautionary steps are taken. It is widely speculated that it is only a matter of time before small groups acquire nuclear weapons; and, we suppose, only a few decades later before individuals of great wealth and power can, if they so choose, be 'armed' with nuclear weapons. These circumstances are generally recognized as unstable.

We may, of course, destroy ourselves. Alternatively, it is possible that we will make fundamental changes in the way we manage the planet, in the way we organize our society, in what passes for conventional political, economic, social and ethical judgements so as to ensure *with extremely high reliability* that nuclear weapons do not pose a significant threat to the human species. The suggestion that such fundamental changes can be instituted is often considered 'Utopian', and different people have different estimates of how difficult it would really be – given the alternative – to make such changes. In a widely quoted remark, Einstein said that the advent of nuclear weapons has changed everything except our way of thinking, when it is precisely our way of thinking that must change if we are to survive nuclear weapons. It seems very plausible that, if we do not change our way of thinking, the lifetime of our technical civilization will be short.

If every civilization that invents weapons of mass destruction must deal with comparable problems, then we have an additional principle of universal applicability. Weapons of mass destruction force upon every emerging society a behavioral discontinuity: if they were not aggressive they probably would not have developed such weapons; if they do not quickly learn how to control that aggression they rapidly self-destruct. Those civilizations devoted to territoriality and aggression and violent settlement of disputes do not long survive after the development of apocalyptic weapons. Long before they are able to make any significant colonization of the Milky Way they are gone from the galactic stage. Civilizations that do not self-destruct are pre-adapted to live with other groups in mutual respect. This adaptation must apply not only to the average state or individual, but, with very high precision, to every state and every individual within the civilization. Because we are so newly faced with this difficult alternative, because we can so easily feel our own predispositions to territoriality and aggression, and because our culture provides so little encouragement to a planetary perspective, such an alternative organization, especially for a highly technological society, might seem to us at first unlikely. The required changes might take thousands of years or more, if the society does not destroy itself first. They might involve major new departures in rearing the young, in education, in the structuring of adult society, or even in prosthetic or biological intervention. Perhaps–although we consider this unlikely–very few societies succeed in such a programme. In any case, the only societies long-lived enough to perform significant colonization of the Galaxy are precisely those least likely to engage in aggressive galactic imperialism.

We think it possible that the Milky Way Galaxy is teeming with civilizations as far beyond our level of advance as we are beyond the ants, and paying us about as much attention as we pay to the ants. Some subset of moderately advanced civilizations may be engaged in the exploration and colonization of other planetary systems; however, their mere existence makes it highly unlikely that their intentions are benign and their sensitivities about societies at our level of technological adolescence delicate. This self-sorting of the civilizations in the Galaxy necessarily operates on time-scales $\gg 10^9$ yr. Thus, any interstellar civilization with a lifetime approaching the galaxy-crossing time will have long before selected itself away from aggressive designs. We believe that the apparent difficulties posed by the 'Where are they?' question derive partly from a conceptual model of interstellar colonization that is in poor accord with long-term human history, and that ignores the differences between the exploration of the Earth and the exploration of the Galaxy; and partly from an inappropriate and self-contradictory application of recent human history to the circumstances which prevail after the invention of weapons of mass destruction.

Since Tipler refers repeatedly to *Intelligent Life in the Universe*, perhaps we may be excused for quoting a passage from it:

> 'Finding life beyond the Earth – particularly intelligent life . . . – wrenches at our secret hope that Man is the pinnacle of creation, a contention which no other species on our planet can now challenge . . . The discovery of life on some other world will, among many things, be for us a humbling experience . . . In assessing evidence for extraterrestrial life, and in evaluating statistical estimates of the likelihood of extraterrestrial intelligence, we may be at the mercy of our prejudices. At the present time, there is no unambiguous evidence for even simple varieties of extraterrestrial life, although the situation may change in the coming years. There are unconscious factors operating, in the present arguments of both proponents and opponents of extraterrestrial life.'   (Shklovskii & Sagan 1966, p. 22.)

This is an issue of some importance. Many people have an emotional investment in the outcome. The question touches on religious and political matters where predispositions have traditionally played important rôles. But it is abundantly clear from the history of science that no convincing resolution of this issue is likely to come from protracted debates carried on with great passion and sparse data. We have an alternative denied to the medieval scholastics: we are able to experiment. We can organize a scientifically rigorous systematic search for extraterrestrial intelligence using the technology of modern radioastronomy. That is where the energies should be focused of those concerned with the great issue of the existence of other technical civilizations in the cosmos.

References

Davis, K., 1974. The migrations of human populations, *Scientific American*, **231**, 52–65.

Frautschi, S., 1982. Entropy in an expanding universe, *Science*, **217**, 593–599.

Gurney, W. S. C. & Nisbet, R. M., 1975. The regulation of inhomogenous populations, *Journal of Theoretical Biology*, **52**, 441–457.

Jones, E., 1978. Interstellar colonization. *Journal of the British Interplanetary Society*, **31**, 103–107.

McEvedy, C. & Jones, R., 1978. *Atlas of World Population History*, Penguin, Harmondsworth.

Newman, W. I. & Sagan, C., 1981. Galactic civilizations: population dynamics and interstellar diffusion, *Icarus*, **46**, 293–327.

Sagan, C., 1973. On the detectivity of advanced galactic civilizations, *Icarus*, **19**, 350–352.

Shklovskii, I. S. & Sagan, C., 1966. *Intelligent Life in the Universe*, Holden-Day, San Francisco.

Sivard, R. L., 1982. *World Military and Social Expenditures*, World Priorities, Washington.

Tang, T. B., 1982. Fermi paradox and C.E.T.I., *Journal of the British Interplanetary Society*, **35**, 236–240.

Tipler, F. J., 1980. Extraterrestrial intelligent beings do not exist, *Quarterly Journal of the Royal Astronomical Society*, **21**, 267–281.

Tipler, F. J., 1981a. A brief history of the extraterrestrial intelligence concept, *Quarterly Journal of the Royal Astronomical Society*, **22**, 133–145.

Tipler, F. J., 1981b. Additional remarks on extraterrestrial intelligence, *Quarterly Journal of the Royal Astronomical Society*, **22**, 279–292.

Tipler, F. J., 1981c. Extraterrestrial intelligent beings do not exist, *Physics Today*, **34**. 9–81, *inter alia*.

Tipler, F. J., 1981d. The most advanced civilization in the galaxy is ours, *Mercury*, **11**, 5–37, *inter alia*.

Tipler, F. J., 1982a. Response to letters, *Physics Today*, **35**, 26–38.

Tipler, F. J., 1982b. Anthropic-principle arguments against steady-state cosmological theories, *Observatory*, **102**, 36–39.

PART 5

Detectability and decipherability

SETI science is based on a number of assumptions: that ETs exist, that they are intelligent, that they have a science roughly comparable to our own, that – in principle – their discourse could be intelligible to us. As we have seen, there are weighty reasons in favor of each of these assumptions, but there are reasons against each of them as well. Further assumptions are needed, however, if we are to think that someday we will be discoursing with aliens. For we will have to assume that an extraterrestrial message will be detectable by our instruments, and that we will be able to recognize it when it comes. Beyond this there is perhaps the most important assumption of all, that we will be able to decode and understand the message's contents. The authors in this section examine what reasons we have for making these assumptions.

It is currently thought that an alien message will most likely come in the form of radio signals. Other means of transmission have been proposed, for example gravity waves, tachyons (hypothetical faster-than-light particles), even optical signals. But by and large the assumption – another assumption! – has been that the aliens will beam radio messages at us. The massive extent to which we use radio waves today may be one of the reasons why we think this. A hundred years ago, however, we didn't use them at all, and a hundred years from now we may use them no longer: even today a large amount of television programming goes by cable and fiber optics. For communication across interstellar distances, however, radio waves are an ideal means of transmission. They can be sent across vast distances cheaply, and they can contain large amounts of information.

To receive an interstellar message we have to do more than point an antenna at the sky and listen, for the sky is already full of radio noise, some of it man-made, some of it natural. The science of radioastronomy was born, in fact, when in the late 1920s and early 30s Karl Jansky of the Bell Telephone Laboratories in New Jersey was trying to pinpoint the source of interference to international radiotelephone circuits. While he had identified numerous terrestrial sources – ranging from thunderstorms to man-made interference – he also noticed a hiss at 20.53 megahertz which traveled regularly across the sky. When he plotted the signal against a star map, he found that the source was the center of the Milky Way Galaxy.

Jill Tarter, a research astronomer at the University of California, Berkeley and at the NASA Ames Research Center in Mountain View, California explains the various problems confronting systematic searches of the sky for signs of alien signals. Where do we search? On what frequencies? How will we separate the signal from the background noise against which it will be heard? By using the analogy of searching for 'a needle in a cosmic haystack,' Tarter shows how these and other problems can be solved. She also describes some of the searches already done, as well as the false alarms that have come in – signals mistakenly identified as alien messages.

Tarter has participated in many searches herself, and she gives us some idea of what doing an actual search is like. She concludes with a tabulation of all the systematic searches ever done, an archive of SETI observations from Frank Drake's Project Ozma to those which continue today.

When pulsars and quasars were being discovered in the 1960s, their signals were sometimes taken for intelligent messages. The radiation from pulsars are extremely regular, for example that from Vulpecula, the first pulsar to be discovered, comes every 1.33728 seconds. With rhythmical beats like this coming in from a variety of sources in the universe, how will we ever be able to separate an intelligent signal from all the others? The answer supplied by cryptologist Cipher A. Deavours is that the signal will be nonrandom, but not as regular as naturally produced pulsations. Rather, an intelligent signal will vary according to rules, resulting in patterns that can be identified through standard cryptanalytic procedures.

But identifying an intelligent message is only half the problem: the other half is decoding it, extracting the message contents from the medium which contains it. Deavours gives us reason to be hopeful of eventual success. After all, centuries of analytic techniques have been perfected by codebreakers, and these strategies have enabled cryptologists to decipher all kinds of earthly messages that had been intentionally encoded to prevent disclosure of their contents. Alien communications, moreover, will be encoded in a kind of 'anticryptography,' a code constructed not to prevent disclosure of message contents, but to facilitate it.

Cipher Deavours* is Professor of Mathematics at Kean College of New Jersey, and a founder and editor of the scholarly journal *CRYPTOLOGIA: A Journal of Mathematical Cryptology*. His article in this volume, 'Extraterrestrial communication: A cryptologic perspective,' has a focus different from that of Marvin Minsky's piece, although both are concerned with the intelligibility of aliens. Minsky wanted to show why aliens could be expected to develop a science and grammatical structures much like our own, irrespective of whether those aliens will ever try to communicate with us. Deavours, on the other hand, assumes what Minsky tries to prove and goes on to concern himself with the practical, nuts-and-bolts issues of the cryptologic strategies that will enable us to understand the messages we may receive.

Whatever codes the others might use to communicate with us, there is one language that has been worked out on earth for our use in communicating with them. It is Hans Freudenthal's LINCOS, or 'Lingua Cosmica,' a language for cosmic discourse.

Freudenthal, a Dutch mathematician, worked out the essentials of this cosmic language in painstaking detail. The short excerpts from his book reprinted here give only the barest hint of the scope, depth, and difficulty of his project. His book starts with conventions for the presentation of mathematics. It ends with 'A short proof of Lorentz contraction,' and the 'Relativistic mass increase formula,' both given in the Lingua Cosmica he invented. Will representations of these formulas be broadcast, or received, by us one day?

* You might think (as I first did) that the name 'Cipher A. Deavours' is a hoax: an anagram, perhaps, a fitting joke concealing the cryptologist-author's true identity. But the name is real. Deavours' father, a mathematician, named his son 'Cipher' on purpose.

Searching for extraterrestrials

JILL TARTER

Introduction and overview

Searching for extraterrestrials, specifically for evidence of their technology, has been likened to 'looking for a needle in a haystack': unfortunately, the SETI is not that easy! To understand why, let us pursue the haystack analogy a bit farther. Suppose that you were (for whatever reason) firmly convinced that there was a needle of extraordinary intrinsic value buried in a haystack, how would you seek it out? You might wish to explore every haystack everywhere, but since you are of limited longevity (although boundless in your enthusiasm) this would inevitably mean that you could spend only a trivial amount of time investigating any particular haystack. The earth consists of 148.9 million sq. km (1.5×10^{10} ha) of solid surface, of which 0.73×10^{9} ha is potentially arable and 22% or 1.57×10^{8} ha is currently under cultivation. Assuming that hay-growing utilises 5% of this land and that each hectare devoted to hay cultivation yields on average 50 haystacks of conical dimensions 1.8 m in diameter and 1.5 m in height, then in a productive working lifetime of 50 years, you will be able to spend a little less than 4 s examining each haystack, even if you could travel from one to the next instantaneously. Such an approach might yield positive results, if it could be implemented, but there is another temporal aspect thus far neglected. Hay is stacked at least once a year (in some places more frequently than that) and, after it has dried, the haystack slowly disappears as the hay is put to the use for which it was grown. Therefore, if you are to find the needle you seek (assuming that it has indeed been buried within this year's hay harvest), you must complete your exhaustive haystack search within a fraction of a year, before the hay and needle are digested by some cow and subsequently transformed into a steak or returned to the earth for recycling or eventual weathering to destruction. So in fact you really have only 0.05 s per haystack for your search. Nevertheless, this is more time than 0 s per haystack, and has a greater probability of a successful outcome than if you never looked at all. There is, however, an entirely different way to proceed.

If you are dissatisfied with the trivially short time available to search through an individual haystack when all haystacks are searched, you might decide to ask if some haystacks have a greater *a priori* probability than others of containing the needle you seek. In this way you could limit the subset of all haystacks which you actually search and spend longer on each one, without appreciably decreasing your probability of success. In fact, if you could say for sure that the needle you are looking for had to be placed in the haystack by a member of the indigenous population, and further that there are certain areas on this planet where hay is grown, but needles are unknown, then a restricted search strategy, which excluded haystacks from such locales, would have a higher probability of success, because more time could be spent on each of the remaining haystacks which had the possibility of containing the needle. Indeed, you could push this strategy to an extreme whereby you used logic and your preconceived ideas of the nature of the needle and the motivations behind its being placed in a haystack, until you narrowed down the number of haystacks to be searched to that number which could just be explored, with the degree of thoroughness you defined as necessary and sufficient, within the time available to you. If all your logic and preconceptions turn out to be correct, then you will have remarkably enhanced your probability of success, because you can examine the selected haystacks in much more detail. But if your *a priori* filtering was in error, you may well have excluded precisely that subset of haystacks which contain the needle or needles! Until now, we have implicitly discussed the existence of only one needle, but if one exists maybe there are many. Perhaps the occurrence of these intrinsically valuable needles in haystacks is the rule rather than the exception. In that case, the best search strategy might well be to go to the nearest available haystack, acquire it from the farmer, and subject it to the most exhaustive search that your finite existence permits, so that if a needle is there, you cannot possibly miss it. How do you know when you have looked enough? When can you let the cow devour your haystack?

The answer, in part, depends upon what you are seeking. If you can define precisely the properties of this intrinsically valuable needle, then you can enumerate a series of search strategies sufficient to locate the needle on the basis of at least one of its properties. But 'needle' is a generic term and covers items made of metal, wood, plastic, bone, stone and other materials. Some have 'eyes' or holes near the pointed end, some at the opposite end, some have no 'eye' but rather a hook (as in crochet needle) or a cap or two pointed ends (as in knitting needles). Some have solid shafts and some are hollow. Some are larger than a haystack (as is the Seattle Space Needle) and some are extremely small (as is a phonograph cartridge needle). Some are straight and some are curved and some are both, depending on their chemical and thermal environment. In the end it is only possible to state what classes of needles you wish to define as desirable and to perform a series of investigations optimised for the detection of those precise classes. If you use a magnet to search for a steel needle, you will not find the one made of platinum; however, if they are large enough, both may be found by X-raying the haystack.

With all of the various strategies for finding needles in haystacks alluded to above, an unsuccessful search cannot culminate in the statement that there are no needles of intrinsic value buried in haystacks. It is only possible to state that, at the times and places searched, no needles of a certain specific type were found. So it is with searches for extraterrestrial intelligence. Positive results from such ventures may be questioned, debated and acted upon in various ways by various individuals and political entities, but negative results do not prove non-existence.

It is quite proper, for those who consider the detection of an extraterrestrial species to be a 'needle' of intrinsic value, to proceed with strategies which attempt to explore, in a systematic way, as much as possible of the multi-dimensional phase space that represents the 'Cosmic Haystack.' And it is incumbent upon those searchers to report, with detailed caveats, the negative results of their efforts. Over the past 25 years, a number of astronomers have done just that. Individuals have whittled down the overwhelming generalized search problem to smaller ones that they could handle in exactly the ways illustrated above with the terrestrial needle and haystack. Each of the circumscriptive variations mentioned has been tried at one time or another and many more will be attempted in the future, until success is achieved or the exploration has been so exhaustive as to rule out the possibility of all *conceivable* evidence. What we cannot conceive must be left to the realm of the unexplored. Such is the nature of needles and haystacks and intelligent species.

In spite of having rather heavily worked over the needle and haystack analogy, we have still not demonstrated its intrinsic weakness in describing the problems encountered in SETI: although the haystack search is an enormous undertaking, it is still manageable! An exhaustive search of all haystacks everywhere is indeed beyond the capabilities of one individual, but it is certainly feasible if enough individuals are motivated to participate. Except to say that billions and billions of individual pairs of human eyes have watched the skies over the last 2 million years, without preserving any verifiable proof of the existence of extraterrestrial intelligences (a fact that some scientists use to rule out the possibility of optical laser transmissions between other civilizations, see e.g. Connes 1982), the search for evidence of extraterrestrial technology is not likely to benefit from a significant multiplexing of individual efforts. The number of potential ways to search is vast, and most search strategies will require some form of sophisticated technology that inevitably imposes a severe limitation on the rate at which various hypotheses can be tested. Resources devoted to any search of this nature will probably remain modest, so that *time* is the only answer to conducting a systematic search. Human history certainly records instances of projects which have lasted through centuries of time, but these endeavors were generally motivated by a strong and enduring organizational structure (such as a religion/ church). SETI may be the one example of a long-term project (which necessarily grows at any stage on the basis of previous failures) undertaken and continued out of individual curiosity. Although the needle and haystack analogy fails to convey the true scale of a search for extraterrestrial intelligence, the schemes and limiting strategies being pursued at any one instant do indeed resemble the haystack problem in the different ways illustrated above.

What to search for

How *do* you seek evidence of extraterrestrial intelligence? It depends very much upon who you think is smarter. Until the nineteenth century, when we began to use electromagnetic radiation for communication on other than a very local point-to-point basis, and the twentieth century, when we began to get a measure for the true size and age of the universe around us, most philosophers and scientists who accepted the probability of life beyond the earth ascribed to it a level of intelligence and technology comparable or perhaps somewhat inferior to our own. This was also comfortably consistent with at least the Christian concept of Man (and presumably extraterrestrial Man) being created in the image of God. Therefore the onus of contact fell upon us. Many schemes were devised, whereby we could signal our existence and intellectual prowess to our neighbors nearby (with particular attention being given to the Martians). Karl Gauss suggested the planting of forests and fields giving graphical proof of the Pythagorean theorem, while Joseph von Littrow wanted to ignite kerosene-filled ditches in the Sahara desert in an assortment of geometric shapes. At the turn of this century Nikola Tesla strained the electrical power production capacity of Colorado Springs, Colorado, and environs to attempt to send long wavelength radio signals to attract the attention of local extraterrestrials and he became convinced that his signals had been answered (Tesla 1901). Guglielmo Marconi believed he had received long-wavelength radiation from extraterrestrial beings while onboard his experimental communications yacht 'Elettra,' without his even having tried to signal them (Marconi 1921). Although the very long wavelength involved precludes the possibility of the 'Elettra' signals being extraterrestrial, this incident probably marks the first instance of a search that was serendipitous to the development of the field of radio communications and eventually radio astronomy.

The 1920 Shapley–Curtis debate before the National Academy of Sciences over the existence of 'island universes' far outside of our own Milky Way galaxy was finally settled when Edwin Hubble (1924) published the first of a series of corrections to the distance scale with which we size our universe. Hubble's later report of the expansion and large finite age of the universe (Hubble & Humason 1931) and the growing acceptance of Darwin's (1859) theory of evolution finally humbled scientist and philosopher alike. Evolution may have produced intelligent species on planets orbiting other stars, but they are probably not at all like us, and the chances are that they are a lot older and consequently more advanced. Therefore the onus of contact is on them, and we should search by receiving rather than sending.

And just what should we be prepared to receive? If our wildest fantasies and tales of science fiction do in fact reflect the reality of more advanced civilizations, then we should expect them to land in a shining craft on the lawn of the White House or the Kremlin or even on top of Devil's Tower and proceed to tell us everything we ever wanted to know. In this case, the search consists of waiting and attempting to arrange for a hospitable greeting committee. On the other hand, if the universe is teeming with beautiful worlds of water, populated by the super intelligent analogs of terrestrial dolphins and whales, lacking manipulative organs and any technology, then no passive search strategy will detect them directly. Their presence will only be deduced, if at all, by inference from an observed non-equilibrium chemistry in the biosphere of those planets at such future time as our astronomical instruments are of sufficient resolution and power as to isolate and image the individual members of distant solar systems. Or when we have fulfilled our own science fiction fantasies and can travel there ourselves in a shining craft.

It seems to me that our science fiction projections are far too restrictive and probably do not in any way correspond to the realities of an advanced extraterrestrial civilization. Nevertheless, I think the arrival of 'shining craft', theirs or ours, is not a good bet. The bottom line is energy. If you wish to transmit a single bit of information, it requires far less energy to send a relatively long wavelength photon than to send any particle possessing mass. The kinetic energy of an electron traveling at half the speed of light is 10 orders of magnitude greater than the energy of a microwave photon, and it takes twice as long to get there! Since there seems to be no way around the speed of light limitation, and since journeys requiring apparent onboard durations less than standard lifetimes (for the convenience of any biological passengers) are prohibitive in their energy requirements, some researchers have concluded that the only matter to make the long treks between interstellar civilizations will be automated probes. And indeed a few searches of our local solar environs tailored to look for probes or other artifacts have been conducted, and more will undoubtedly be promoted as we resume our scientific exploration of the solar system in the coming decades. Still other researchers have concluded from these same energy arguments that the only thing that will ever cross the vast interstellar distances will be information. If this information is intentionally transmitted for reception at a distance (as opposed to being leakage from transmissions intended for any civilization's local consumption), then the carrier of information we should be prepared to search for will:

(1) have a minimum energy per transmitted quantum;
(2) have a velocity as close to c as possible;
(3) suffer negligible absorption and deflection by the interstellar medium;
(4) be easy to generate, modulate and capture.

Item (3) argues against any charge-bearing particle that would suffer deflection against the diffuse intergalactic magnetic field. I have already pointed out that (1) and (2) argue for the use of a zero (or near zero) mass particle, while (4) suggests that the appropriate particle is a photon. Various forms of neutrinos, gravitinos and other 'inos' have been proposed for communications between advanced civilizations. These particles should propagate with almost no loss through the interstellar medium, because they possess very tiny cross-sections for interaction. But for precisely this reason, they are extraordinarily difficult to capture and (at least for us) to generate and modulate. If this is indeed the communication channel of choice for advanced civilizations, our search strategy must be to become more advanced ourselves, because at the moment we cannot play in that league. However, there remains the possibility of using electromagnetic radiation (photons) for long-distance communication. We do so ourselves routinely, and the majority of the searches performed to date or proposed for the near future have concentrated on photon detection.

But which photons? Photons can propagate across long distances if their frequencies (energies) lie somewhere within the range of 10^4 Hz to 10^{24} Hz. To be easily recognizable or detectable by the receiving civilization, they must be received with a number density which significantly exceeds the number density of similar photons received from natural astrophysical phenomena in the universe. In other words, nature provides a background noise pollution (the bread and butter of astronomers everywhere) above which the sender must 'shout' in order to be heard by the receiver. It makes good cosmic sense to 'shout' at frequencies where the sky is naturally quietest, in order to conserve energy, and it has been argued that this is precisely what communicating extraterrestrial civilizations will do. If the sender does not know the precise location of the receiver, or is broadcasting for the benefit of a number of widely separated receivers, then the microwave region of the spectrum is to be preferred on these grounds. However, if the relative positions of sender and receiver are precisely known, then a case can be made for infrared or optical frequencies purely on the basis of detected signal to noise ratio for a given transmitted power. Our own detector technology is currently far superior at microwave frequencies as compared with the infrared and optical, so it is not surprising that most of the searches carried out to date have concentrated on this region of the spectrum. Even so, previous searches have only scratched the surface of what might be required in a systematic search for evidence of microwave signals of extraterrestrial intelligent origin.

The past 25 years

The year 1984 means many things to many people. To SETI enthusiasts, it marks the quarter-century anniversary of the publication of the paper by Morrison & Cocconi (1959), which first suggested using existing radio telescopes to search for evidence of signals from extraterrestrial intelligences at or near the 21 cm line emission frequency of neutral hydrogen. In many ways this was the first step in the continuing process of gaining scientific credibility for the SETI endeavor. As a result, in these past 25 years, various individuals (mostly radio astronomers) have felt confident enough to approach bureaucratic scheduling committees world-wide with formal proposals for use of specific pieces of instrumentation for the express purpose of searching for extraterrestrial intelligence. Perhaps the most astonishing result is that quite often they succeeded in acquiring the requested time! In appreciation of the enormity of the search task, other individuals have themselves constructed SETI-specific detectors and installed them in amateur-dedicated observatories,* the rationale being that there may well be very strong signals capable of being detected by relatively simple instruments within the grasp of individuals or small groups. Such signals may thus far have failed to be detected because there are so many possible combinations of signal frequency, direction of propagation and signal type that other more sensitive searches, being extremely constrained by available observation time, have not yet happened to look in the right places at the right times in the right way. An amateur-dedicated SETI observatory just might succeed because there would be a great deal more time to investigate a large number of combinations for strong signals. Still other individuals have wondered whether sensitive astronomical instruments might not have already recorded evidence of the existence of extraterrestrial intelligence, but have ignored it in an attempt to cull out the astrophysically relevant data they were seeking. Where possible, these individuals have reanalyzed pre-existing data in a manner intended to discover specific types of signals which might have gone unnoticed.

Below, we shall look at some of the details and amusing occurrences associated with the three types of observing programs outlined above: directed, dedicated and parasitic. The appendix to this chapter (p. 192) contains a brief archive of more than three dozen SETI programs known to the author. It is unfortunately true that the publication of negative results is not very esthetically pleasing nor necessarily rewarded, and thus the number of references citable to these searches is limited.

Directed searches

Project OZMA, the very first SETI program, was performed by Frank Drake in 1959–60 at the same time that Morrison and Cocconi were collaborating on their historic paper (see reference *1* in appendix). These observations required several hundred hours of telescope time. They involved pointing the newly constructed 26 m antenna of the National Radio Astronomy Observatory (NRAO) at the two closest solar-type stars, Tau Ceti (a G-type dwarf star like the sun at a distance of 11.9 light years) and Epsilon Eridani (a K-type dwarf star at 10.7 light years), and 'listening' with a special stable radiometer built by Drake and collaborators especially for OZMA. Drake hoped to detect a signal falling

---

* Observatories constructed by radio 'hams' to search for ET signals and to do that exclusively.

within a single narrow 100 Hz channel as he scanned it across a 400 kHz band of frequencies centered on 1420.405 MHz (the 21 cm line frequency at which atomic hydrogen, the most abundant element in the universe, radiates naturally). In this very first search, Drake and colleagues confronted almost all of the problems inherent in this type of endeavor and in addition drew attention to the special role of the nearly monochromatic or narrowband carrier wave (CW) type of signal for the purposes of intra-species communication and quite possibly for inter-species communication as well. The reason Drake had to build a special radiometer for SETI, even though he was working at the pre-eminent NRAO, is that he was trying to detect a type of signal that we routinely produce ourselves, but as far as we know, *Mother Nature never does!* Thus all the available radioastronomy receiver systems were (and continue to be) very insensitive to this form of signal. Indeed when the radioastronomers do become cognizant of the presence of a narrowband signal within their system, they inevitably say some very unprintable remarks and attribute the signal to radio frequency interference (RFI) from a local terrestrial source. RFI is indeed a problem for astronomers and SETI folk alike. Since, relatively speaking, it is produced right next door (or flying overhead), it is likely to be much stronger than the weak astrophysical or ETI signals originating at intergalactic or interstellar distances. Drake used two separate horns at the telescope focus, one of which looked at the star of interest, and the second of which looked at a patch of the sky just to the side of the star; the signals coming out of the two horns were then subtracted. In this way Drake tried to eliminate the effects of interference from natural and man-made causes. A signal which was present in the output of both horns had nothing to do with the star, and would tend to cancel out, on average, during the subtraction. Even so, after an unsuccessful observing run on Tau Ceti, the radiometer indicated a positive detection almost as soon as it was pointed to Epsilon Eridani! During the time that this signal continued to be detected, Drake fought against his own growing excitement, and attempted to perform in a coolly rational manner those real-time tests available to him in order to confirm that the signal was actually emanating from the star or some other source at great distance in that direction of the sky. The signal did not persist and was never again reacquired from the direction of that star; the best guess is that it had its origin in classified airborne equipment. This experience illustrates two very important aspects of the SETI enterprise: the benefit arising from real-time signal detection and analysis protocols, and the intrinsic conflict between the public/popular interest in the search and the requirement to verify and independently confirm the reality of any candidate detection (implying a certain minimum period of silence and restraint on the part of the investigators). But scientists are human, and recently many people without specific disciplined training have joined the search with clever search strategies based on home components and computers. We should not be surprised if a large number of 'detections' are reported in the media, and we should not grow too cynical about all the false alarms. In 1965, the *New York Times* reported that some Russian scientists had announced the detection of a very advanced technological civilization (see reference *20* in the appendix). Actually they had discovered that the quasar CTA102 possessed an unexpectedly complex spectrum, a matter of non-trivial interest in its own right! It would be a shame if the 'cry wolf' phenomenon caused us to ignore a valid detection of any sort. The corollary of course is that we will be required to investigate invalid detections of all sorts!

Many of the observational programs I personally have been involved with have acquired data at too high a rate to allow the luxury of real-time analysis procedures. In those cases, data analysis becomes more like detective work. What in the world caused the signal that was recorded on magnetic tape and picked out by the computer so many months later? Such deduction can be exciting, frustrating, and a certain percentage of the time inconclusive. Thus most of us in this business have a drawer somewhere containing a list of 'birdy' coordinates and frequencies that we reobserve at every opportunity, hoping to reproduce a prior detection. Some of these exercises in amateur sleuthing have left me with a profound respect for the difficulties inherent in any search that presumes to operate within the environment of human technology and yet attempts to be sensitive to a range of signal types which might be indicative of an extraterrestrial technology. In 1977, Tom Clark, Jeff Cuzzi, David Black and I recorded hundreds of magnetic tapes in one 5-day observing period using the large 300 ft (91 m) antenna of the NRAO (see reference *12* in the appendix). The data on the tapes contained information about signals from the direction of several hundred nearby solar-type stars at frequencies near the natural frequency of the 18 cm emission line of the hydroxyl radical, which occurs in many concentrations of gas and dust surrounding stars and in between. (Since hydrogen and the hydroxyl radical are the dissociation products of water, it has been suggested that these two emission line frequencies serve as natural signposts to mark a region of the microwave spectrum, the so-called 'waterhole', which might be of particular significance for communicating water-based species.) The lack of any real-time data analysis capability conspired with the specific characteristics of the instrumentation in use to produce two tantalizing detections, and a memorable increase in adrenalin levels. The 300 ft (91 m) telescope is steerable in only one coordinate, which means it is a transit instrument. Any given source can be observed for a few minutes only, and then cannot be observed again until the rotation of the earth causes it to pass overhead at about the same time the next day. While conducting OZMA II at NRAO, a few weeks before we arrived, two other radioastronomers, Pat Palmer and Ben Zuckerman (see reference 6 in the appendix) had compiled a list of about a dozen stars which had yielded possible signals at the 21 cm frequency. They were kind enough to leave this 'hit list' for us and we added these stars to our own large list of stars to be observed at 18 cm. Since these stars were potentially special, we decided to observe them every day when they passed overhead. Over a year later, when the computer at NASA's Ames Research Center had finished processing all of our data tapes, we discovered that every single day that we had observed one particular star from this special list, we had detected very strong signals! We had made standard precautionary observations in the vicinity of every star to identify RFI events, but did not find the signals in our 'off' source tapes and therefore were inclined to associate the detected signals with the star itself. To be sure, we detected signals at a number of discrete frequencies each day and, at least in the reference frame of the star, they were not the same set of discrete frequencies each day. *But* we never saw such a pattern of signals when we looked at any other star. I am certain we would have spent a lot more time investigating these signals if Dr Clark had not himself been a radio ham. He noticed that the pattern of frequencies we were detecting mimicked the frequency pattern with which CB radio channels were allocated. During our short observing run, another thing had been happening each day when the star transited: at just about 8 a.m. EST, the telescope operators changed shift. And in Greenbank, West Virginia (where the observatory is located to shield it from intereference by large commercial transmitters) the pickup trucks and cars of most of the telescope operators were equipped with CB radios! Some extra remedial efforts on one of our 'off' source reference observation tapes eventually confirmed that we had also detected these signals when pointing the telescope near, but not on the star on one day only. The 'off' source measurements were necessarily separated in time by about 6 min from the 'on' source measurements, and this slight time difference and Murphy's law had worked to our disadvantage. The other false alarm produced by the Greenbank observations was more intriguing than believable. In a paper entitled, 'Through the Looking Glass', presented at the 1976 International Radio Science Union meeting in Amherst, Massachusetts, Dr Thomas Gold of Cornell University had reasoned that an extraterrestrial civilization might take advantage of the opportunity afforded by the existence of interstellar masers* to amplify greatly (at no energy cost to the transmitting society) an interstellar beacon. In a reprinted version of this talk, which I received around the time of these observations, Gold suggested that the transmitting society might broadcast a narrowband signal at a frequency close to that of one of the natural molecular masers and in the direction of a nearby maser cloud. Anybody on the other side of the cloud and along the initial line of sight of the transmission, would be treated to an intensely strong and obviously artificial signal whenever their astronomers got around to studying the maser cloud. While at Greenbank, we studied a number of known hydroxyl maser clouds in an attempt to verify the performance of our novel observing technique for achieving extremely high frequency resolution, and to see if these objects held any unexpected surprises when dissected into so many narrow frequency components. In one of these sources we found a very strong signal of less than 5 Hz bandwidth at a frequency just slightly higher than the highest frequency maser feature we could detect. Exactly what Gold had predicted! Unfortunately for SETI, this particular signal at its particular frequency turned out to be an artifact of our tape recorder and data acquisition techniques and turned up over and over again.

Not all searches have concentrated on looking at nearby solar-type stars. Some on dedicated systems (described below) have attempted to survey the entire sky, while others have accommodated observing time limitations by attempting to define another limited set of special directions on the sky. The galactic center is one obviously special point of reference for any Milky Way civilization. The stellar density here is very high, but so is the incidence of very energetic and violent phenomena, the kind of location astronomers might classify as 'A nice place to visit, but who would want to live there?' Indeed, perhaps nobody can live there, but it has been suggested that an advanced technology might choose to erect an artifact or a beacon there, where it is sure to be stumbled upon in the course of any other civilization's astronomical research. Steve Lord and collaborators have expanded upon this idea and investigated a strip of the sky that lies along the axis of rotation of the galaxy (including the galactic center as one of the search elements). For pragmatic as well as philosophical reasons (see the 1981 entry in the appendix) these investigators conducted their search at yet another 'magic frequency,' the

---

* In a maser, concentrations of interstellar molecules can coherently amplify microwave radiation incident upon them in the same way that atoms can intensify optical radiation in lasers.

115 GHz emission line of the abundant molecule carbon monoxide. Another limited set of observational directions on the sky lies much closer to home and is related to the possibility of artificial probes being sent over large interstellar distances. Suppose that you could target a distant planet which had the potential for supporting some form of life. You might have the technology to send an automated probe to that planet in the hope of discovering and informing any indigenous civilization. But planets exist long before intelligent species arise (if the earth is a typical example) and your probe might arrive too early to catch the natives. It might well be desirable to park such a probe in a stable orbit around the planet, and wait for intelligence to evolve on it. Near earth there are a small collection of such neutrally stable orbits in the vicinity of the 'Libration points,' i.e. those places where the combined effects of gravitational forces from either the earth and the moon or the earth and the sun just cancel. In 1979 and again in 1981 (see references *18* and *23* in appendix), Frank Valdes and Robert Freitas used optical telescopes to examine these regions of the sky searching for reflected sunlight from objects in such parking orbits. The telescopes used were not large, and their negative results so far exclude only shiny objects many kilometers in length.

In 1983 these last two observers also performed an experiment to detect indirectly the presence of extraterrestrial nuclear technology. One inevitable by-product of nuclear fusion of hydrogen to helium is tritium. This isotope of hydrogen has a natural half-life of about 12.6 years, and would therefore not be expected to occur naturally in space. Valdes and Freitas hypothesized that another civilization that is fusing hydrogen as an energy source might create a detectable level of tritium line emission at 1516 MHz. They used an 85 ft (26 m) radio telescope (see 1983 entry in appendix) to search for this short-lived isotope in the vicinity of nearby stars. Their failure to detect any tritium emission is more difficult than most observations to assess, since the mean density of tritium 'likely' to be released during H fusion is a complete unknown and may well be zero. A cynical evaluation might conclude that if it were detectable, then they would no longer be around to converse with us.

Dedicated searches

As mentioned previously, the scope of the SETI problem is so vast, that time is one of the major assets in making a systematic search. Telescopes built to do astronomy at any wavelength are typically oversubscribed and available only occasionally, for short periods. Further they are instrumented to do astronomy, and therefore far from ideally instrumented for SETI. The Cyclops Study (Oliver & Billingham 1973) concluded that it would be desirable to build a large dedicated array of 100 m radio dishes in order to enhance the probability of detection of an ETI signal. Perhaps such a facility must necessarily be built some day, but it probably will not be built on or near the surface of the earth because of the ever increasing density of RFI from our own usage of the spectrum for communication purposes. Telescopes dedicated for SETI today tend to fall into two categories; those constructed for astronomical observations or commerical applications, which have lost their source of operational funding and tend to be outmoded, and those that have been constructed specifically for SETI by enthusiastic amateurs and tend to be small.

In 1973 the Ohio State–Ohio Wesleyan University Radio Observatory (OSURO) began to conduct a full-time SETI sky survey (see reference 8 in appendix). This meridian transit instrument had completed the astronomical survey work for which it had been constructed and funded by the National Science Foundation, and financial support had ceased. The antenna, looking like a giant metallic football field, provides the equivalent collecting area of a 53 m parabolic dish and has access to 60% of the sky. It has operated daily in an automated fashion as a volunteer effort under the direction of John Krauss and Robert Dixon, two of the professional astronomers who helped to build it. The receiver system and data acquisition software are quite modest by modern observatory standards and have been designed to search specifically for signals transmitted at 1420.405 MHz in the rest frame of the galaxy. A great deal of care has been taken so that the limited amount of real-time data processing which can be accomplished during its unattended operation will make the system as insensitive as possible to RFI and tolerant of the many natural sources of 21 cm radiation that pass through the beam daily. Nevertheless, the equipment continues to record a number of candidate signals that pass all the established criteria for an ETI signal, and a certain portion of the telescope time and human energy must be devoted to reobservation and attempts to confirm these 'birdies.' This ultimately imposes a real operating cost, and the directors have been quite creative and energetic in their attempts to finance the ongoing effort. The magazine *Cosmic Search* was conceived as a potential funding source for this venture and nearly induced bankruptcy before it was turned over to an independent and profit-oriented organization. During its brief existence it did provide a vehicle for discussing the 'Wow!' signal detected in August 1977 and never seen again. (This signal was so unexpectedly strong that the student who found it while routinely scanning the observation summaries wrote 'Wow!' on the computer printout). I mention this for two reasons: there currently exists no proper forum for the scientific examination of interesting possible signals, and while any systematic survey must rely on silicon processors to analyse the datastream in real-time, human expertise must be immediately available to invoke whatever verification procedures seem to be appropriate while the extraordinary signal is still accessible. In spite of its shoestring operation and limited sophistication, there seems to be a deep base of public support for this endeavor. When recent real estate dealings by the two controlling universities threatened to sell the land out from under this facility to make room for a golf course, there was an overwhelming reaction from scientific and lay communities alike which appears to have embarrassed the principals into submission. Barring any new legal setbacks, this facility will probably continue its sky survey work for some time to come.

It is difficult to assess what long-term impact the recent public interest and individual participation in SETI programs will have, unlike the situation at OSURO. Although I stated at the outset of this chapter that it was unlikely that SETI could benefit significantly from the multiplexing of individual efforts, there have been in recent years some highly visible exceptions in the arena of small dedicated SETI facilities. Most efforts have not yet reached the point where they might be included in the appendix of SETI observations, but there appears to be a small army of amateurs in Silicon Valley (and elsewhere) who feel that the best possible use for a home computer is as a SETI signal analyzer! One example of this type of involvement is a project called AMSETI being sponsored by Delta Vee Inc., of Santa Clara, California (a new role for the old Viking Fund organizers). These amateurs meet regularly and have a column in the Delta Vee magazine *Astrosearch*. They are currently being advised by scientists involved in the NASA SETI program and other local signal processing experts on the techniques for constructing low noise amplifiers, large Fast Fourier Transforms to run on microcomputers, and all the other goodies required to turn a garden variety antenna built for satellite TV reception into a dedicated SETI observatory. It is far too soon to tell how long this enthusiasm will last and how many data of relevance will ever get analyzed by this and similar groups worldwide. There has been more than one author writing about SETI over the last 25 years who has opined that a signal will arrive only after our own leakage TV and military radar signals have been detected by a neighboring intelligence, and that this return signal will acknowledge receipt in such a way and with such power that it will not take very sophisticated equipment to hear it. If this scenario is correct, then at least for the past few years and for the immediate future, there are plenty of eager listeners.

I shall close this section on dedicated systems with a brief discussion of what is truly the Cadillac of such endeavors to date: the Sentinel project conducted by Paul Horowitz with financial support from the Planetary

Society. This is one particular case where it can be said that 'less is more,' less bandwidth that is. If you are convinced that you know exactly at what frequency a signal will be broadcast, and further that it will certainly take the form of a purely monochromatic carrier wave and will be transmitted from there to here in a way that minimizes any tendency for that frequency to change or drift in time, then you would build a single channel receiver with extremely narrow bandwidth. In this way you limit the addition of noise from other nearby frequencies at which the signal does not exist. This is what is often called the 'magic frequency' game. Since the space between the stars is not completely empty, photons travelling over long interstellar distances bump into things along the way (so-called multipath scattering) and therefore a signal which was generated as a completely pure tone at one frequency will always arrive spread over a narrow band of frequencies. This sets a lower limit to how narrow you should make your receiver channel bandwidth so as not to exclude any of the signal. And in practice (unless you have some additional inside information) you really need to use more than one channel because 'there' and 'here' are not precisely enough defined. Rather the transmitter and receiver are moving through space relative to one another and, unless you can establish *a priori* which is the proper reference frame for defining the magic frequency, it will be necessary to search over a small range of frequencies corresponding to a small range of velocities (earth around the sun, both earth and sun around the center of mass of their mutual motion, etc.).

'Suitcase SETI' is just such a receiver; it searches 2 orthogonal polarizations with 128000 channels each only 0.03 Hz wide, so it can look for carrier wave signals in a band 2 kHz wide centered on the magic frequency. So much care must be taken to compensate for the rotation of the earth about its axis during the time that the data are being acquired, that there is an additional curious advantage from a system of this type: it is very easy to recognize and discard signals that are the result of local RFI, for they end up spread across many many channels, whereas the particular type of ETI signal being sought will occupy one or at most two adjacent frequency channels. This type of receiver has now been installed on the old unused Aggassiz 26 m radiotelescope belonging to Harvard University, and like Ohio State this sytem scans the area that passes overhead each day in an automated and largely unattended mode. Each year it will scan 68% of the sky at two different magic frequencies and is likely to remain reliable and in use for quite some time. If you believe in the validity of this type of search strategy, then Project Sentinel is about as good a job as it is possible to do with state-of-the-art technology, and this piece of the Cosmic Haystack should soon be well explored. The problem with magic frequencies is that there are as many magic frequencies as there are individuals who propose them (and their ranks are swelling).

Parasitic searches

I have mentioned in several contexts the problems posed by limited funding and access to instrumental facilities. In a number of different parasitic searches that have been conducted to date, individuals have attempted to circumvent these difficulties by reusing astronomical data already 'bought and paid for' or by sharing the telescope while an astronomical observation is in progress. A good example of the former approach is a study of 25 globular clusters by Chip Cohen and colleagues (see reference *15* in appendix). High quality data at the frequencies associated with water and hydroxyl masers (1.3 and 18 cm) had been collected at a number of radio observatories in order to study the frequency of occurrence of these masers in globular clusters. Globular clusters contain about 10^5 stars, most of which are far older than the sun, and not generally regarded as good candidates for extraterrestrial life sites. (Early in the history of the galaxy there was a paucity of heavy elements from which to condense rocky planets; multiple generations of stars had to manufacture and distribute them.) However, evidence from masers and other arguments now indicate that these clusters may have also hosted more recent binges of star formation and thereby produced an opportunity for extraterrestrial habitation. In each case, the entire globular cluster was observed within the beam of the telescope providing 10^5 simultaneous opportunities to test Gold's hypothesis about interstellar amplifiers, or just the general contention of hydroxyl radical and water molecule magic frequencies. These old data were re-examined with search algorithms optimized to uncover types of signals distinctly different from those of astronomical origin and earlier interest.

The second approach is exemplified by an ongoing search for regularly pulsed signals with periods ranging from 0.3 to 1.5 s, which has been carried out with the 100 m Efflesberg telescope of the Max Planck Institute für Radioastronomie (see 1977 Wielebinski & Seiradakis entry in appendix). The telescope is usually pointed at supernova remnants, HII regions or along spiral arms, any place where pulsars are likely to be found. However, the pulse-seeking software works just as well if the antenna happens to be pointed at a nearby solar-type star, and it would dutifully report the presence of pulses of the type to which it was sensitive, whether or not the pulses were of intelligent or natural origin. During the course of the regularly scheduled observing program, it is often possible to fill in a gap of time before the next source rises above the horizon by observing a star in the part of the sky currently visible. But it is also possible to take this kind of approach one step farther, and to argue that in general the radioastronomers will *always* be using the telescope to observe a region of the sky and a frequency which are of potential interest to SETI experimenters. Stuart Bowyer and his colleagues have constructed a little black box called Serendip (see reference *11* in appendix), which sits in the corner of the observing lab of the 85 ft (26 m) telescope at Hat Creek, California, and automatically scans the data being collected by astronomers for other purposes. This sytem works by tapping-off some of the available voltage from the telescope front end before it ever gets to the observers' signal detection equipment. A microcomputer steps a 100 channel spectrum analyzer across the entire frequency band available, and at every step it decides whether the spectrum contains an interesting relatively narrow ETI signal in addition to the relatively broad astrophysical signal being studied. If such a signal exists and if it exceeds a pragmatically set threshold level, the spectral data are written onto a magnetic tape and saved for later processing. The processing of these tapes takes much more time than it took to compute the spectra that are written on them, and that is one major disadvantage of such parasitic techniques. The coordinate positions for the telescope and the frequencies of the observation must first be identified by means of the time recorded by an internal clock. Then a sequence of different analyses must be performed to look for patterns or clustering in the detected signals when correlated with sky coordinates, time of day, day of week, location of astrophysical sources, baseband or laboratory frequencies and a number of other parameters which have proven useful in identifying (after the fact) which signals are of interest and which can be attributed to what type of interference. There is usually a substantial residue of questionable cases to be added to the list of candidates awaiting reobservation and verification at such time as SETI can be scheduled as the primary user of the telescope.

In the spirit of 'whatever is available,' SETI observations have even been performed when for all other practical purposes, the telescope was shut down for repairs. Sam Gulkis and his colleagues (see 1983 entry in the appendix) took advantage of the fact that the large 64 m telescope at NASA's Deep Space Network tracking station in Tidbinbilla, Australia, had to be decommissioned for several months in order to repair a structural fault in the cement pedestal of the telescope mount. All of the front end and receiver hardware remained in place, the telescope could in fact collect data, it simply could not be moved. For most purposes such an immobile antenna would have been useless, but stowed in a position slightly off the zenith, it could in fact observe a limited range of declinations over and over again as the sky passed overhead. Most of the sky is fair game for a SETI program: the nearest stars are uniformly distributed over a sphere and the more distant stars, being concentrated towards the central plane of our flattened Milky Way galaxy, are visible for part of each day. With financial aid from the Planetary Society, Gulkis *et al.* put together a microprocessor-based spectrum analyser with 256 channels and a software protocol for recognizing and recording signals of sufficient strength to be real and not just noise. These several months' worth of data were later analyzed in an attempt to identify signals which repeatedly arose from the same general direction on the sky. Since the other large telescopes in NASA's network are plagued with this same structural fault, it is likely that this system will see further use in the future.

The next 25 years

The reader who has gone this far may well be sufficiently impressed (or bored) with the details and diversity of the sample of previous observing programs given in the last section to ask 'what is left to be done?' The answer is: *everything*! Although we have argued that the microwave region is a good place to look for evidence of extraterrestrial technology, we must be reminded that it is not necessarily the only place. And even if it were, these past 25 years of individual efforts (albeit sometimes very clever and/or Herculean) have not even scratched the surface. This is another often used metaphor, like the needle and haystack, which completely fails to impart the correct scale of the matter. Some numbers are required. It is possible to make a minimum number of assumptions and thus come up with a model for the Cosmic Haystack (Tarter & Zuckerman 1980), whose volume in multi-dimensional search space is calculable. In the same way the various fractional 'volumes' of the Haystack which have been explored by each of the searches in the appendix can be calculated and added together. To date we have only investigated 10^{-17} i.e. 1/100 000 000 000 000 000 of the volume of the Cosmic Haystack! This leaves a little something left over for the next 25 years.

Using the same measurement scheme, we can evaluate the bi-modal search strategy that has been proposed by NASA as a systematic SETI exploration of the microwave region (Gulkis *et al.* 1980 and Wolfe *et al.* 1981). This five-year observing program will improve on what has already been done by a factor of 10^7, and seems a reasonable next goal! However, the NASA Microwave Observing Program (MOP) cannot promise success (no program can) because this still means that only 10^{-10} of the Cosmic Haystack will have been explored. The hope is that it will be the right portion. As mentioned at the outset, there are many guesses as to where in the Haystack a signal may be found, and each search strategy attempts to optimize for that guess, but it is still a guess, and we may be investigating the wrong part of the Haystack. Investigations of other parts of the Haystack or even of other Haystacks must await new technologies.

What is it that NASA plans to do, and what kinds of signals might it find? First, it will spend the next four or five years developing the tools that will be required and learning how to do the job, since nobody has tried anything quite like this before. Then with the help of the newly designed very special purpose multichannel spectrum analyzers having in excess of 10 million separate channels (some as narrow as 1 Hz) and signal processors capable of keeping up with the real-time data rate of 10 billion bits of information each second, the MOP will use existing large radio telescopes throughout the world to listen for signals of many different types, over a large range of frequencies while pointed at about 800 nearby stars like our own sun and while driving the antennas rapidly across the sky to cover all possible directions: a targeted search and an all sky survey. The targeted search achieves great sensitivity by looking at a small number of probable directions for a long time and covers the frequency range from 1 GHz to 3 GHz. It will probably require a few per cent of the time available on the three or four largest telescopes for a five-year period. The sky survey sacrifices sensitivity by looking in a very large number of directions for a short period of time, but it does investigate the possibility of intrinsically strong signals originating from directions other than those picked as likely (perhaps the transmitting civilization lives far away around a star too faint to appear in any of our catalogs, or perhaps the signal comes from a direction not associated with a star). The sky survey will make heavy use of the 34 m antennas which are part of NASA's tracking network and which, because of a decline in planetary missions, are available for a large percentage of the time.

What kinds of signals could these two searches detect? Neither search will be able to eavesdrop on the ETI's TV traffic even to the distance of the nearest stars, unless they allocate more power to their transmitters than we do. However, the targeted search should be able to pick up signals, whose power level is no greater than has been estimated for an orbiting solar power station, if they originated within 30 light years of the sun. The earth's most powerful transmitter is the Planetary Radar focused by the 1000 ft (305 m) antenna at Arecibo, Puerto Rico and used to bounce signals off of Mars and Venus to map their topography. The sky survey could detect the extraterrestrial analog of this transmitter out to 30 light years, while the targeted search should see it all the way out to 300 light years. Detection at greater distances requires the technology and motivation to construct more powerful transmitters than is our current capacity, but this seems like a good bet.

And if this is not enough? No doubt other regions of the spectrum will be explored as photon detector technology matures. It has been suggested by Mumma and Bogan (Eberhart 1983) that other civilizations might erect giant mirrors in synchronous orbit above a planet whose atmosphere contained a weak natural carbon dioxide laser. The resulting cavity might produce a modulated and detectable signal in the infrared at 10.4 microns even if the planet and star could not be resolved by a distant observer. The infrared has also been suggested by Dyson (1960) as the place to look for evidence of large-scale astro-engineering projects. An advanced civiliz- ation might erect a nearly complete spherical shell around its star near the outer edge of its planetary system in order to trap and utilise all the available stellar radiation for power. The backside of such a shell would glow in the infrared like a few hundred degree Kelvin black body. Depending on how much of the primary starlight escapes, this so-called Dyson sphere would be visible as a cool infrared point source or as an unusual infrared excess in the spectrum of an otherwise normal star. Some work on this hypothesis has already begun from ground-based observatories (see 1980 Witteborn entry in appendix), but the millions of infrared sources soon to become available when the Infrared Astronomi- cal Satellite (IRAS) catalog is completed will provide fertile ground for future systematic investigations.

When and if we finally achieve the technical competence and financial motivation to send automated probes to distances hundreds and thousands of astronomical units (1 AU = the mean distance between the earth and sun) beyond the sun, it is possible that we might explore this region looking for the focal spots in the gravitational lens system of our sun where distant signals at wavelengths of less than 1 mm may be amplified and highly detectable (Eshleman 1979).

But there is also a lot more that can be done in the microwave region. The largest gains to be made here are sensitivity and frequency coverage, and both require leaving the surface of this planet. Our own atmosphere produces an increasingly loud background noise as frequencies are in- creased above 10 GHz and portions of the submillimeter and far infrared are completely absorbed by the atmosphere. Going into space and working from a permanent space platform provides access to otherwise inaccessible portions of the spectrum, but low earth orbit does not solve (and may in fact exacerbate) the problem of terrestrial RFI. Studies are currently underway to determine whether a large space antenna can effectively be shielded from the strong electromagnetic field produced by ground based transmitters and orbiting satellites. In this context, the communications satellites become the analog to terrestrial transmitting towers right next door, but there are no valleys in low earth orbit in which to hide the antennas. One feasible solution is to shield the antenna with the solid body of the moon, placing it (or them in Cyclops fashion) on the far side. The economics are prohibitive from a SETI consideration alone, but may become manageable as a part of an otherwise motivated lunar settlement. Lastly, an antenna constructed in low earth orbit could potentially be solar sailed out to trans-lunar distances, where shielding becomes a more tractable problem.

Will any of this be done? It is very difficult to assess the public support for advanced technology, which is completely justifiable from a scientific point of view, but actually becomes necessary only as the result of previous failures. It is to be anticipated that although ETI signals may not have been detected, the new technology developed at each stage may well have led to some serendipitous discoveries in more traditional disciplines and therefore further advances may well be supported.

The next 25 years may not bring us a definitive answer in our quest for extraterrestrial intelligence, but there is a related problem whose solution is within reach during that period, and is beginning right now. This is the search for extrasolar planetary systems. One assumption implicit in SETI is that warm, wet rocky planets are plentiful in our galaxy and elsewhere. There is no universally accepted model of how our own planetary system formed, but most theories of star formation predict that planets are the rule and not the exception. In the recent data from the Infrared Astronomical Satellite, we may have even observed the first evidence of dusty disks around young stars, from which a planetary system may someday condense. But today we have no firm evidence of any other planetary system anywhere. In the next decade ground-based and space-based systems will achieve the capacity for indirectly detecting the presence of Jupiter-sized planets (and later on, earth-sized planets) around other stars because of the sinuous path they cause the stars to follow as they wander through the skies. When two bodies are in orbit, both orbit around their common center of mass. If one body is massive and the other is not, then the center of mass will lie very close to the massive body (in the case of the earth and sun, the center of mass lies deep within the sun itself). When viewed from a distance, this mutual dance of star and planet around the center of mass will cause a slight wobble to be visible in the apparently straight track of the massive star across the sky. Discerning this wobble requires extremely precise measurements over many years and is greatly facilitated if the stellar twinkling caused by our own atmosphere can be avoided. Soon we shall know whether the planetary link in the logical chain that produces SETI enthusiasts is indeed sound. Soon we may have a better list of stars at which to point our antennas.

Almost all of what has been said to this point involves SETI programs conducted in the United States or by Americans. This is not meant to imply that SETI is not conducted by other nations; indeed there is a strong base of international interest in this field. I have presented details from observational programs with which I am most familiar and therefore introduced a strong bias. Since SETI is so very much a global or species concern rather than a national issue, I hope that the next 25 years will produce something that has not happened in the past 25: joint cooperative observing programs.

I wish to thank the reference librarians at the Berkeley, California, Public Library and the University of California Berkeley, Library of Natural Resources, who helped me to count the haystacks. Also thanks to Peter Backus and Kent Cullers, colleagues at NASA Ames Research Center whose critical comments kept this somewhat non-traditional treatment of SETI within bounds. This work was completed under NASA Cooperative Agreement NCC-2-36 with the University of California Berkeley.

References

Connes, P. (1982). Paper presented at IAF Congress in Paris, France.
Darwin, C. (1859). *Origin of the Species*. London: J. Murray.
Dyson, F. (1960). Search for artificial stellar sources of infrared radiation. *Science*, **131**, 1667.
Eberhart, J. (1983). *Science News*, **124**, 181.
Eshleman, V. R. (1979). Gravitational lens of the sun. *Science*, **205**, 1133–5.
Gulkis, S., Olsen, E. & Tarter, J. (1980). A bi-modal search strategy for SETI. In *Strategies for the Search for Life in the Universe*, ed. M. D. Papagiannis, pp. 93–105. Boston: Reidel.
Hubble, E. (1924). N.G.C.6822, a remote stellar system. *Astrophysical Journal*, **62**, 409.
Hubble, E. & Humason, M. (1931). Velocity distance relationships among extra-galactic nebulae. *Astrophysical Journal*, **74**, 43–80.
Marconi, G. (1921). *New York Times*, Sept. 3, 4:4.
Morrison, P. & Cocconi, G. (1959). Searching for interstellar communications. *Nature*, **184**, 844–6.

Oliver, B. M. & Billingham, J. B. (1973). *Project Cyclops Report*, NASA CR114445.

Tarter, J. C. & Zuckerman, B. (1980). Microwave searches in the U.S.A. and Canada. In *Strategies for the Search for Life in the Universe*, ed. M. D. Papagiannis, pp. 81–92. Boston: Reidel.

Tesla, N. (1901). *Collier's Weekly*, **26**, no. 19, 4.

Wolfe, J. H., Edelson, R. E., Billingham, J., Crow, R. B., Gulkis, S., Olsen, E. T., Oliver, B. M., Peterson, A. M., Seeger, C. L. & Tarter, J. C. (1981). The search for extraterrestrial intelligence: plans and rationale. In *Life in the Universe*, ed. J. Billingham, pp. 391–417. Cambridge, Mass.: MIT Press.

APPENDIX

Archive of SETI observing programs 1959–84

Summary of SETI observing programs (Feb. 1984)

Date	Observer	Site	Instr. size (m)	Search freq. (MHz)	Freq. resol. (Hz)	Objects	Flux limits (W/m^2)	Total hours	Comments	Ref.
1960	Drake 'OZMA'	NRAO	26	1420–1420.4	100	2 stars	4.E − 22**	400	Single-channel receiver.	1
1963	Kardashev & Sholomitskii	Crimea deep space station		920	10 MHz	Quasar			Reported detection of CTA102 as possible Type III civilization.	20
1966	Kellermann	CSIRO	64	Many, between 350 & 5000	Full bandwidth for each feed	1 galaxy	±0.5 FU	—	No 'notch' of ETI origin was detected in galaxy 1934–63.	2
1968 & 1969	Troitskii, Gershtein, Starodubtsev, Rakhlin	Zimenkie, USSR	13	926–928 & 1421–1423	13	12 stars	2.E − 21**	11	25 Channels with F = 13 Hz were spaced 4 kHz apart: coverage not continuous.	3
1968 & on	Troitskii	Gorky	Dipole	21 cm 50 cm 1 m	—	All sky search		Cont.	Search over all sky visible by single dipole.	24
1970 & on	Troitskii, Bondar, Starodubtsev	Gorky, Crimea, Murmansk, Ussuri	Dipole	1863 & 927 600	—	All sky search for sporadic pulses	1.E + 4 FU	700 & cont. at 50 % time	Network of isotropic detectors: cross correlation from 2 or 4 sites over 8000 km.	4
1970 to 1972	Slysh	Nancay	40 × 240	1667	20 kHz	10 nearest stars				
1971 & 1972	Verschuur 'OZPA'	NRAO	91 & 43	1419.8–1421 & 1410–1430	490 & 6900	9 stars	5.E − 24 & 2.E − 23	13	384 channel correlator on-line.	5
1972	Kardashev & Steinberg	Caucausus, Pamir Kamchatka, Mars probe		40–500	2 MHz	Omni-directional			'Eavesdropping' search for pulses.	
1972 to 1976	Palmer, Zuckerman 'OZMA II'	NRAO	91	1413–1425 & 1420.1–1420.7	6.4×10^4 & 4000	674 stars	1.E − 23**	500	384 channel correlator on-line.	6

Date	Observer	Site	Antenna	Frequency (MHz)	Bandwidth	Object	Flux limit	Integration	Remarks	Ref.
1972 & on	Kardashev, Gindilis	Eurasian network	Dipole	1337–1863	—	All sky search for sporadic pulses	1.E + 4FU	—	2 or more sites operating simultaneously.	7
1973 & on	Dixon, Ehman, Raub, Kraus	inst. for cosmic radiation OSURO	53	1420.4 rel. to gal. cen. ±250 kHz	10 & 1 kHz	All sky search	1.5E − 21**	Cont.	Receiver is tuned to hydrogen rest frequency relative to gal. cen. (as a function of direction).	8
1974 to 1976	Bridle, Feldman 'Qui appelle?'	ARO	46	22 235.08 ±5 MHz	3×10^4	70 stars	1.E − 22*g	140	70 solar-type stars within 45 lt yrs have been observed to date.	
1974	Wishnia	Copernicus satellite	1	3.E9	—	3 stars	—	—	Search for UV laser lines.	9
1974 & on	Shvartsman 'Mania'	Zelenchuk-skaya	6	550 nm	$\Delta\Lambda \leq 10^{-6}$ nm	21 peculiar objects	—	—	Optical search for short pulses of length 3×10^{-7} to 300 seconds, and narrow laser lines.	21
1975 & 1976	Drake, Sagan	NAIC	305	1420 & 1667 & 2380 B = 3 MHz	1000	4 galaxies	3.E − 25**	100	Search for type II civilizations in local group galaxies.	10
1975 to 1979	Israel, De Ruiter	WSRT	1500 max. baseline	1415	4×10^6	50 star fields	2.E × 23**	400	Searches of 'cleaned' maps prepared for the WSRT background survey, looked for positional coincidence between residual signals and AGK2 stars.	
1976 & on	Bowyer et al. U. C. Berkeley 'Serendip'	HCRO	26	1410–1430 & 1653–1673	2500	All sky survey	5.E × 22**	—	Automated survey parasitic to radio astronomical observations.	11
1976	Clark, Black, Cuzzi, Tarter	NRAO	43	8522–8523*	5	4 stars	2.E − 24**	7	VLBI high speed tape recorder combined with software direct Fourier transformation to produce extreme frequency resolution (non-real time).	

(*continued*)

Summary of SETI observing programs (Feb. 1984) continued

Date	Observer	Site	Instr. size (m)	Search freq. (MHz)	Freq. resol. (Hz)	Objects	Flux limits (W/m^2)	Total hours	Comments	Ref.
1977	Black, Clark, Cuzzi, Tarter	NRAO	91	1665–1667*	5	200 stars	1.E − 24**	100	VLBI high speed tape recorder combined with software direct Fourier transformation to produce extreme frequency resolution (non-real time).	12
1977	Drake, Stull	NAIC	305	1664–1668*	0.5	6 stars	1.E − 26**	10	VLBI high speed tape recorder combined with optical processor to produce extreme frequency resolution (non-real time).	13
1977 & on	Wielebinski, Seiradakis	MPIFR	100	1420	20 000 000	3 stars	4.E − 23	9	Candidate stars are inserted into ongoing program which searches for pulsed signals with periods of 0.3 to 1.5 s.	14
1978	Horowitz	NAIC	305	1420 ± 500 Hz	0.015	185 stars	8.E − 28**	80	Assumes that signal frequency was corrected at the source to arrive at rest in heliocentric or barycentric laboratory frame.	15
1978	Cohen, Malkan, Dickey	NAIC / HRO / CSIRO	305 / 36 / 63	1665 + 1667 / 22 235.08 / 1612.231	9500 / 65 000 / 4500	25 globular clusters	1.8E − 25 / 1.1E − 22 / 1.5E − 24**	40 / 20 / 20	Passive search for Type II & III civilizations using astronomical data originally observed to detect H$_2$O and OH masers in globular clusters.	16
1978	Knowles, Sullivan	NAIC	305	130–500 (spot)*	1	2 stars	2.E − 24**	5	Attempted 'eavesdropping' using MKI VLBI tapes as in Black *et al.* 1977.	

										Ref.
1979	Cole, Ekers	CSIRO	64	5000 ± 5 MHz and ± 1 MHz	10^7 and 10^6	Nearby F, G and K stars	$4 \times 10^{-18**}$	50	Simultaneous pulsed events in both 2 MHz and 10 MHz filters are sought in detectors having time resolution of 4 μs.	17
1979	Freitas, Valdes	Leuschner Observatory UCB	0.76	550 nm	—	Stable 'halo orbits' about L4 and L5 libration points in earth–moon system	$10 \leq m_V \leq 19$	30	Attempt to discover evidence of discrete objects (such as interstellar probes) in stable orbits about L4, L5 by study of 90 photographic plates.	18
1979 & on	JPL & UCB Serendip II	DSS 14	64	S & X band B = 10 MHz	19 500	Apparent positions of NASA spacecraft	$8.E-24**$	400 to date	Automated survey parasitic to spacecraft tracking operations using 512 channel autocorrelator and 100 channel correlator with microprocessor control.	
1979 to 1981	Tarter, Clark, Duquet, Lesyna	NAIC	305	1420.4 ± 2 MHz & 1666 ± 2 MHz	5 & 600	200 stars	$1.E-25**$	100	Rapid 1-bit sampler and high speed tape recorder run in parallel with 1008 channel correlator. Software direct Fourier transformation as in Black et al. 1977.	22
1980	Witteborn	NASA–U of A Mt Lemon	1.5	8.5–13.5	1	20 stars	N magnitude excess 1.7	50	Search for IR excess due to Dyson spheres around solar type stars. Target stars were chosen because too faint for spectral type.	
1981	Lord, O'Dea	U. Mass	14	115 000	20 000, 125 000, 4×10^8	North galactic rotation axis = 5°–90°	$1.E-21**$	50	Search for signals near $J = 1 \rightarrow 0$ CO line frequency from a transmitter somewhere along the Galactic rotation axis.	19
1981	Israel, Tarter	WSRT	3000 max. baseline	1420	4×10^6, 10×10^6	85 star fields	$8.E-22$ to $6.E-24$	600	Parasitic search similar to Israel and De Ruiter using 'uncleaned' maps stored at Gröningen and Leiden, and AGK3 catalog.	

(continued)

Summary of SETI observing programs (Feb. 1984) continued

Date	Observer	Site	Instr. size (m)	Search freq. (MHz)	Freq. resol. (Hz)	Objects	Flux limits (W/m²)	Total hours	Comments	Ref.
1981 & on	Biraud, Tarter	Nancay	40 × 240	1665–1667	97.5	300 stars	$1.E-23$**	80 to date	8-level 1024 channel auto-correlator with stepped first LO to extend frequency coverage at modest resolution.	
1981	Shostak, Tarter	WSRT	3000 max. baseline	1420.4 rel. to gal. cen. B = 156 KHz	1200	Galactic center	$1.E-24$**	4	Use of interferometer to search for pulsed signals from galactic center in range of periods from 40 s to 2 h.	
1981 to 1982	Valdes, Freitas	KPNO	0.61	5500	—	Earth-moon L1 thru L5 sun-earth L1, L2	$10 \leq m_v \leq 19$	70	Attempt to see discrete artifacts (few m in size) in stable orbits near Lagrange points. Studied 137 IIIaF photographic plates.	23
1982	Horowitz, Teague, Linscott, Chen, Backus	NAIC 'Suitcase SETI'	305	2840.8 B = 4KHz and 1420.4 B = 2KHz	0.03 1-linear / 0.03 2-circular	250 stars / 150 stars	$4.E-26$** / $6.E-28$**	75	Dual 64K channel real time microprocessor based spectrum analyzer with video archiving and swept LO frequency to test 'magic frequencies.'	
1982	Vallee, Simard-Normandin	ARO	46	10 522	185 MHz	Galactic center meridian	$1.E-19$**	72	Search for strongly polarized signals by mapping field 1/4° × 25° along b = 0°	
1983	Horowitz 'Sentinel'	Oak Ridge (Harvard)	26	1420.4 and 1665.4 and 1667.3 and 2840.8	0.03 Dual circular B = 2 KHz	Sky survey	$8.E-26$**	Cont.	'Suitcase SETI' as the back end of automated sky survey at 4 or 5 'magic frequencies' over a 5-year observing period.	

Year	Observer	Site		Frequency	Bandwidth	Sky survey	Sensitivity	Hours	Description	
1983	Damashek	NRAO	92	390 ± 8	2×10^6	Sky survey (pulsars)	1.E − 28	700	16 MHz sampled at 60 HZ; 8 contiguous frequency channels. Search for single dispersed pulses and telemetry (bit-stream) signals.	
1983	Valdes, Freitas	HCRO	26	1516 ± 2.5	4.9 KHz 76	80 stars 12 nearby stars	3.E − 24**	100	Search for radioactive tritium line from nuclear fusion.	25
1983 & on	Gulkis	DSS 43	64	8 GHz and 2380 ± 5 MHz	40 KHz	Partial southern sky	2.E − 25 2.E − 22**	800 & on	Sky survey of constant declination strips (3 from −28.9 to −34.3 by April 1983) whenever antenna stowed.	
1983 ?	Gray	Small SETI obs.	4	1419.5– 1421.5	1–3	?	1.E − 22	Cont.	Dedicated search system constructed by amateurs. (Scheduled open 1983.)	

*These experiments corrected frequencies for the motion of the observed stars with respect to the Local Standard of Rest.

**Quoted sensitivities refer to signal/noise ratio = 1.

B = bandwidth; m_V = visual magnitude; FU = Flux Unit.

References to Appendix

1. Drake, F. D. (1960), *Sky and Telescope*, **39**, 140
2. Kellermann, K. I. (1966). *Australian Journal of Physics*, **19**, 195.
3. Troitskii, V. S., Starodubtsev, A. M., Gershtein, L. I. & Rakhlin, V. L. (1971). Soviet *Astronomy Journal*, **15**, 508.
4. Troitskii, V. S., Bondar, L. N. & Starodubtsev, A. M. (1975). *Soviet Physics Uspekni*, **17**, 607.
5. Verschurr, G. L. (1973). *Icarus*, **19**, 329.
6. Palmer, P. & Zuckermann, B. (1972). *The NRAO observer. 13*, No. 6, 26. Sheaffer, R. (1977). *Spaceflight*, **19** (9), 307.
7. Belitsky, B., Lawton, A. & Gatland, K. (1978). *Spaceflight*, **20**, 193.
8. Dixon, R. S. & Cole, D. M. (1977). *Icarus*, **30**, 267. Kraus, J. D. (1979). *Cosmic Search*, **1**, No. 3, 32.
9. Morrison, P. (1975). Letter to directors of Radio Observatories dated 29 August 1975 which appears in NASA SP-419, p. 204.
10. Sagan, C. & Drake, F. (1974). *Scientific American*, **232**, 80.
11. Turner, G. (1975). Langley, D. (1976) and Gilbert, B. (1977), unpublished theses for MS degree in Department of Computer Science and Electrical Engineering at UCB. Murray, B., Gulkis, S. & Edelson, R. E. (1978). *Science*, **199**, 485.
12. Tarter, J., Black, D., Cuzzi, J. & Clark, T. (1980). *Icarus*, **42**, 136.
13. Tarter, J., Cuzzi, J., Black, D., Clark, T., Stull, M. & Drake, F. (1979). To be published in *Acta Astronautica*, paper 79-A-43 presented at 30th IAF Congress in Munich, Germany.
14. Horowitz, P. (1978). *Science*, **201**, 733.
15. Cohen, N., Malkan, M. & Dickey, J. (1980). *Icarus*, **41**, 198.
16. Sullivan, W. T. 3rd, Brown, S. & Wetherill, C. (1978). *Science*, **199**, 377.
17. Cole, T. N. & Ekers, R. D. (1979). *Proceedings of the Australian Society of Astronomy*, **3**, 328.
18. Freitas, R. A. & Valdes, F. (1980). *Icarus*, **42**, 442.
19. Tarter, J. & Israel, F. P. (1981). Paper IAA-81-299 presented at IAF Congress Rome, Italy, September 1981.
20. Kardashev, N. S. (1964). *Soviet Astronomy Journal*, **8**, 217. Sholomitskii, G. B., IAU Information Bulletin on Variable Stars, 27 February 1965. *New York Times*, editorial, 13 April 1965, p. 36.
21. Shvartsman, V. F. (1977). *Communication of the Special Astrophysical Observatory*, no. 19, p. 39.
22. Tarter, J. C., Clark, T. A., Duquet, R. & Lesyna, L. (1983). Paper No. IAA-82-263 presented at IAF Congress Paris, France, October 1982: to be published in *Acta Astronautica*.
23. Valdes, F. & Freitas, R. A. Jr (1983). *Icarus*, **53**, 453.
24. Interview in *Leninradskaya Pravda*, 2 November 1982.
25. Kuiper, T. & Gulkis, S. (1983). *Planetary Report*, **3**, 17.

Site abbreviations

ARO	Algonquin Radio Observatory, Ontario, Canada
CSIRO	Commonwealth Scientific and Industrial Research Organization, Epping, New South Wales, Australia
	NASA Deep Space Network:
DSS 14	Goldstone, California
DSS 43	Tidbinbilla, Australia
HCRO	Hat Creek Radio Observatory, Castel, California
HRO	Haystack Radio Observatory, Westford, Massachussetts
KPNO	Kitt Peak National Observatory, Tucson, Arizona
MPIFR	Max Planck Institut für Radioastronomie, Effelesberg, West Germany
NAIC	National Astronomy and Ionospheric Center – Arecibo Observatory, Arecibo, Puerto Rico
Nancay	Observatoire de Nancay, Nancay, France
NASA	National Aeronautics and Space Administration
NRAO	National Radio Astronomy Observatory: Greenbank, West Virginia Tucson, Arizona Socorro, New Mexico
OSURO	Ohio State – Ohio Wesleyan University Radio Observatory, Columbus, Ohio
U of A	University of Arizona
UCB	University of California Berkeley
U. Mass.	Five College Radio Astronomy Obervatory, Amherst, Massachusetts
WSRT	Westerbork Synthesis Radio Telescope, Westerbork, The Netherlands

Extraterrestrial communication: a cryptologic perspective

CIPHER A. DEAVOURS

The press photo shows clearly the three men crouched in deep study over the paper strips on the table before them. All three are in shirtsleeves because August of 1924 found the Washington, D.C., area immersed in one of its usual summertime excesses of heat, humidity and stale air. At the center of the trio, a young man holds a magnifying glass in one hand and a pen in the other. This gentleman, though unknown to the general public, is one of America's greatest puzzle solvers. William F. Friedman is chief of the US Army's code section and himself a cryptanalyst *par excellence*. Tonight, Friedman is searching for messages from the planet Mars.

Among the astronomers, physicists, and radio engineers who participated in man's first attempt to listen for an extraterrestrial communication, the inclusion of a cryptologist can only be attributed to farsightedness on the part of the project's organisers. Cryptologic practitioners occupy themselves with the rather curious pursuit of keeping secrets for their employers while trying to discover those of others. The science of constructing and breaking secret codes and ciphers is to many people a mystery. While other scientists observe nature and seek simple models to explain what they have seen, cryptanalysts deal with problems deliberately made as difficult to resolve as human intellect will allow. Amazingly enough, they often succeed.

A cryptologically trained person could be of use to scientists probing the cosmos for signals from other worlds in two areas: first, in determining whether or not a valid communication had been received and, secondly, in its subsequent decipherment and interpretation. Most people realize only the second use.

It seems likely that scanning the electromagnetic spectrum for extra-terrestrial signals with serious intent is going to involve more than tuning a directional antenna in the neighborhood of 1420 MHz (the radio emission line of hydrogen) and monitoring a few selected areas of the sky. Massive intercept attempts are going to involve significant data processing. There has to be rapid processing of intercepts to exclude signals of terrestrial origin as well as those from errant satellites of domestic origin. As interception attempts are stretched to their limits to hear the faintest of signals, cosmic 'noise' becomes a factor in extracting the wheat from the chaff. Could we recognize a message from another world even if we managed to intercept one?

A cryptologist is, perhaps, uniquely qualified to aid in these problems. Day-to-day analysis of cryptographic systems is, in large part, the search for patterns and non-random phenomena of diverse sorts. The analysis must be done quickly so that intercepted text does not pile up. Analysis must often probe cryptographic systems where random looking ciphertext is intermingled with truly meaningless random 'garbage' to fool the unwary and render the solution process difficult.

Once a signal is extracted from the ether, the problem becomes one of decipherment. The term 'anticryptography' has been coined to describe the process of formulating an easy to understand galactic language. The prefix 'anti' is employed because what is wanted is a text which is especially clear to the recipient in contrast to ordinary cryptography which seeks to disguise the message's meaning. The reverse of anti-cryptography is not, however, anticryptanalysis. It is old-fashioned crypt-analysis itself. So, no matter how the alien sender chooses to write his message, our task is cryptanalytical.

We shall assume in the sequel that an intercepted message would most likely be pulse modulated and can be represented as a succession of binary 0s and 1s although the general approach would remain the same in any case. There are reasons to believe that an alien communication might be of this sort. Any creature whose sensory receptors range over a portion of the electromagnetic spectrum would have a sense of brightest/darkest, loudest/softest, hottest/coldest, or, in short, 0 and 1. Further, non-randomness of the intercepted data would be likely to be recognized even if it could not be interpreted. For example, an infrared sensing creature who digitizes a 'picture' from his environment and transmits it can expect the receiver to perceive the non-randomness of the pattern after it has been unscrambled even if the content of the transmission is unclear.

What will an alien message look like? Undoubtedly it will include some clearly indicated non-randomness and error detecting devices such as sending the text in fixed length pulse groups. Clearly, an extraterrestrial civilization would not be content merely to send a distinguishable text such as

101101110111110111111101111111111010110111011111110 . . .

which consists of of the first few prime numbers repeated over and over. While such a communication might serve the purpose of calling attention to its intelligent origin, it would do little else and would not be worth the effort. After all, the purpose of communication is the interchange of information.

We can expect intercepted communications to be voluminous. The sender wants at least some of the transmission to be understood well enough so that a reply can be formulated by the receiver. Would the sender then formulate his message in some artificial interstellar language such as Hans Freudenthal's *LINCOS* (Freudenthal 1960, and this volume)? Possibly parts of the transmission might take this form, but, in the author's view, artificial languages are likely to introduce artificial difficulties for the receiver. Friedman, for instance, thought the mysterious Voynich manuscript to be an early attempt at an artificial language. After decades of study, no one has yet deciphered the text.

More likely, the text will consist of many things: diverse language samples, pictorial or related data, mathematical and logical puzzles, along with some things we have yet to conceive. One of the purposes of this discussion is to argue that this approach would not confuse the receiver but would offer a real chance that the receiver might comprehend the intended meaning of the messages.

One who has not sat before reams of pages containing meaningless looking groups of symbols cannot appreciate what it is to do cryptanalysis, nor what can sometimes be accomplished with it. As an example, consider a traditional type of code system. A large codebook may replace words, phrases, or entire sentences with a single group of letters or figures. Most people would despair at ever making sense of messages coded in such a system, especially if the underlying language were one they did not know. Yet such problems are routinely solved by cryptanalysts. Gradually a codebreaker will recognize patterns in the coded messages. These and other characteristics will enable partial meanings to be established and explored. It is of interest that codes can often be solved where the underlying language of the plaintext is not known for certain. One can also gain an immense knowledge of the structure and character of a communication without understanding a single thought expressed therein. For intergalactic communication, this offers much hope that we may succeed in deciphering what is received.

The problem which we would face in cryptanalyzing an alien communication is not unlike that which archeologists face in the decipherment of 'dead' languages. In this case, no overt attempt has been made to hide the meaning of the text such as in ordinary cryptographic systems. The Rosetta Stone, which contained parallel plaintext in three languages (Greek, Demonic, and Egyptian hieroglyphs), was the eventual key to deciphering hieroglyphs. The final recognition that Linear B was indeed a variety of early Greek was a sheer act of sustained cryptanalysis on the part of Alice Kober and later Michael Ventris, both of whom used traditional cryptologic reasoning and tests (unbeknownst to themselves) to establish the inflectional nature of the script and its relationship to Greek. The Cypriotic syllabary was established by means of reference to a parallel text in Phoenician. The cryptanalytic breakthrough into Persian cuneiform was obtained via shrewd guesses concerning the names of Persian kings which would have been likely to have appeared in the documents at hand. In these and other cases, the initial clues were often proper names, dates, or descriptions of historical events which were known to the decipherers (Pope 1975). It is primarily for this reason that mathematics is often put forward as a kind of universal language which in itself could provide a type of universally understood basis for communication.

The American cryptologist Lambros Callimahos put forward some sample extraterrestrial messages which a sender might use to convey his knowledge of mathematics (Callimahos 1966). A typical text looked as follows:

AA,B; AAA,C; AAAA,D; AAAAA, E; AAAAAA, F; . . .

from which the recipient might deduce that the symbol B represented the same mathematical quantity as AA, C was equivalent to AAA, and so on. Using such representations, the sender could gradually build up a considerable sophistication in his ability to convey mathematical theorems and relations. While this type of transmission is unlikely to result in the type of information most of us would desire (Where are you? What do you look like? Is there a God?), it would form a good beginning. All of this is, of course, predicated on the assumption that mathematics of the type we have constructed here on earth has universal recognizability.

While parallel plaintext has been a great aid in the study of unknown languages (and in the field of cryptanalysis in general), it would seem that no such help could be expected from an alien civilization which is completely ignorant of our existence or customs. The transmission of pictorial equivalents might provide similar aid if they could be deciphered and recognized as such. Also, the mere transmission of the same message in several different systems (languages) would be invaluable for comparative study and would provide a rich source of clues about the nature of the writer even if the meaning of the text could not be discerned.

To indicate the type of work which might be involved in analyzing an intercepted alien communication, we give some hypothetical examples using only the simplest of cryptologic tools.

Suppose the following block of message text has been intercepted and we wish to examine it further for signs of any communication it might carry.

Message #1

```
101100100010101001000101111010010101000010101010 01
010101011001011101011101000010000010101010000010 11
001101000000010101001011101111011111011001001100 00
011001000110000110100101010011000010101101110100 00
110100101010011000110001000001110000110010110011 10
110011011001110100001101001010100110100100010101 11
101111000110000011001011000001100001011000001100 00
010111010111010010100101100111100010000010101010 00
010110011010000000111000000100011010000010000011 1
110100101000010001110110001011100111011001000101 01
011000101100010101111001110001101011000001011011 00
010110000101100010010110101011000000100011010000 0
010000011111011001000101011101010110110000001110 0101
010110011100111100000011101011001110100110000001 00
101101010111000000100011010000110100101010110001 11
100000100101010101101101010101100000011101001010 11
001110011101011110000101100010000010001111101001 000
001110111110000010110110001000001110000110110110 10
000100000111010101101101010000011100010111010010 11
110100001010110000011000101000110100110010000101 11
000100011010000101010100000111001001010110110011 10
```

The text consists of 1050 bits of which 463 (44 %) are 1 bits. While such a string may appear a more or less random collection of 0s and 1s to the eye, mathematical examination shows it to be exceptionally non-random. We can see this in the following way. Suppose Message #1 were indeed a random collection of bits. In this case, about half the text should be 0s and about half should be 1s. While the difference between 44 % and 50 % does not seem particularly large, probability theory tells us that the chances of this deviation occurring in a string of 1050 truly random bits is about 1 time in 10000. The reception of such a message would immediately call for further analysis.

In accessing the non-random quality of bit strings such as the above , it is not only the frequency with which the 0s and 1s occur but also their arrangement or distribution within the string. For example, the string

101010101010101010101010101010101010101010101010101010101010

contains an equal number of zeros and ones but the alternating arrangement of these bits hardly suggests randomness. The cryptanalyst must constantly be on guard not to miss a trick in observing any sort of inherent non-randomness which may yield information.

Now that a message has presented itself for analysis, what do we do? One of the most common of cryptanalytic tests is comparing a ciphertext against itself to find exceptional statistical behavior. One way in which this may be done is by preparing two copies of the ciphertext and, after shifting one copy against the other, counting the number of coincidences which occur between the two character strings. This general approach is deemed the 'Kappa test' in cryptology (Deavours 1977). The beginning of Message #1 shifted against itself, say 14 positions, would look as follows where coincident positions have been marked by asterisks.

```
          1011001000101010010001011110100101010000 . . .
1011001000101010010001011110100101010000101010100010101 . . .
          ** ***   * * *******  ***    ****    *** *  . . .
```

If the bits of the text are arranged in a random distribution we could predict what the 'coincidence rate' should be when we make such a comparison. In this case, 44% of the text are 1 bits and 56% are 0 bits. Thus, when two bits are compared the chance they will coincide is given by

$$0.44 \times 0.44 + 0.56 \times 0.56 = 0.51 \qquad (1)$$

which is a little higher than the 50% to be expected from a random string of bits. The probability that a consecutive string of N bits will all coincide in Message #1 is given by the formula:

$$p = 0.51^N \qquad (2)$$

assuming the distribution of the bits is random.

Armed with these statistics, we can now proceed to compare the ciphertext string against itself for shifts of 1, 2, 3, . . . positions, while examining the coincidence rate at each shift value to see if it agrees with our expected random rate.

At least two of the shift values produce coincidence rates which are exceptional: those at shifts of 145 (0.54) and 130 (0.53). We may inquire as to why this has happened by looking at the coincidence pattern which is shown below for each case. In the tables a '+' indicates that the two bits in the strings compared agreed and a '−' indicates noncoincidence of corresponding bits.

Coincidence chart for shift of 145

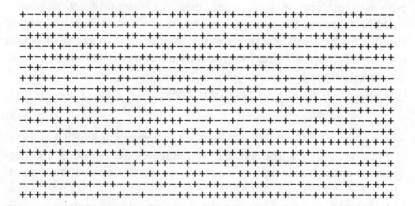

Coincidence chart for shift of 130

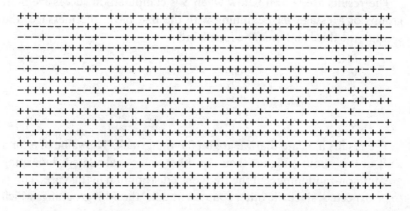

In both cases, the deviation can be seen to arise from exceptionally long strings of coinciding bits. In the first case, 40 consecutive bits agree while, in the second, a 'run' of 37 such bits is found. Equation (2) tells us the probability of these phenomena occurring by chance is on the order of 0.000000000001. Further, it may not be by chance alone that these long repeats are found at shift intervals which were both divisible by 5, since more detailed examination reveals 15 bit repeats at shifts of 70 and 125.

The point is that although we really do not understand what the ciphertext means at this stage, we are able nevertheless to proceed in its analysis.

Since shift intervals of 5 seem preferred places for high coincidence rates to occur, we might test the assumption that the entire text is organized into 5-bit groups. There are only 32 possible 5-bit patterns:

00000, 00001, . . . , 11111. If we take a count of the number of times each 5-bit pattern occurs in the message, then, the resultant frequency count could be examined for non-randomness. We now need some convenient statistic S which will measure the non-randomness of the distribution obtained. Perhaps the most intuitive measure consists of the sum of the 32 frequencies squared, i.e.

$$S = f(00000)^*f(00000) + f(00001)^*f(00001) + \cdots + f(11111)^*f(11111)$$

(3)

where $f(abcde)$ is the frequency the 5-bit pattern $abcde$ which was found in the text. This statistic has the advantage of exaggerating high frequencies and minimizing low frequencies, and serves as a 'measure of roughness' of the frequency distribution. In general, the larger the value of S obtained, the more non-random is the underlying frequency distribution.

The results are shown below when S is computed on successive 5-bit groups of Message #1 beginning with the first bit in the message, the second, the third and then the fourth bit.

Shift	S
0	3000
1	2222
2	2034
3	2134
4	2274

The results indicate that we are on the right track. The value of S for a shift of 0 is about $\frac{1}{3}$ higher than the other values calculated, indicating greater non-randomness of the frequency distribution in this case. This is to be expected, since the other counts 'straddle' the 5-bit groupings which we feel constitute the elementary parts from which the message is constructed. Another way of expressing what the above data show is to say that the message appears considerably less random when viewed from the standpoint of the first statistic.

It is appropriate now to condense the message by representing each 5-bit group by a single character so that it is easier to examine by eye. If we replace each 5-bit group as follows: 00001 = A, 00010 = B, 00011 = C, etc. the text now becomes

```
VHUDKZJPUIJVKUZBAJPKFPAJK
WWVIPLQPZJSAKNPZJSCBAXLVN
YVNPZJSIBWWQPLVAPVAPKUZJK
GBAJPKFPAXBFPHGZJBGLKSVHU
LKBWSQUPKLKALDVUXBFPHGVHU
ZUPGEKGGPGKGIPDVUXBFPZJVG
PIJVZUPGEKGGKXKBAGZHGWPKL
HGAVZBAZVZPNEZKZBVAQHZLPW
       BFPUHGDUVNPFBW
```

Only 26 of the 32 possible 5-bit combinations appear in the text so that the 26 letters of the alphabet are sufficient to represent every 5-bit grouping which occurs. Several longish repetitions in the text have been underscored. At this point, the true statistical and linguistic nature of the text could be examined (assuming much more material were available) for clues to the intended structure and meaning of the message.

Suppose also that another block of text is received directly following Message #1, and similar analysis points to the same type of 5-bit grouping. Using the same substitution as previously, we find the condensed text to read as follows:

Message #2

```
        AVZUPEPUPTHBPZPGKHOYBPHOT
        HPFVZUPAVNGVBZGKAZBDBPTHP
        FVZUPUPXAPKUUBFPTHPFVZUPF
        VWVAZPGVBZDKBZPGHUWKZPUUP
        YVNNPKHYBPWLVAAPMAVHGKHLR
        VHULJHBAVZUPEKBATHVZBLBPA
        PZEKULVAAPMAVHGAVGVDDPAGP
        GYVNNPAVHGEKULVAAVAGKYPHO
        THBAVHGVAZVDDPAGPGPZAPAVH
                WKBGG
```

While the two texts seem dissimilar, the general positioning of the underlined repeats in the second message would suggest that the text of both messages may be the same. If a great deal of text were available, comparisons between the two texts would bring out features of each separate text that would not otherwise be apparent. All this could be accomplished without ever having to understand what the received texts actually say. It is somewhat surprising to learn that most cryptanalysts are not linguists and, in spite of this, are not thereby handicapped to any real extent in solving cryptograms in languages with which they have little acquaintance. For instance, the great French cryptanalyst, Georges Painvain, of World War I fame, solved many complex ciphers of the German General Staff but possessed so little knowledge of German that he was unable to translate the deciphered text after solution. Numerous other examples could be given.

Of course, the plaintext underlying our two sample messages is not of intergalactic origin but a parallel text in English and French. (The well-known passage begins, in French: *Notre Père qui êtes aux Cieux....* Although it may surprise readers unfamiliar with cryptological science, it would have been an easy matter for a trained cryptanalyst to have extracted this message from the number display given as Message #1.

It may be suggested that a true interstellar communication will be in a language so different from those of earth that no analysis of the sort proposed could succeed. As far as is known, all types of communication on this planet (including that of animals) involve what is termed 'redundancy.' By this, we mean that there are non-randomnesses built into the channel of communication. These manifest themselves in terms of irregular frequency characteristics and peculiar arrangements of the written or phonetic units of the message. Redundancy is a mathematical concept and capable of exact measurement given enough samples of the message text. Although redundancy usually exists for 'garble correction,' sometimes physiological and other reasons give rise to it. It would, in fact, be undesirable to remove all redundancy from communication media since there would then be no way to recognize errors in the transmission of information. Written English, for example, is over $\frac{3}{4}$ redundant. This means that, on the average, only 1 out of every 4 letters in written English really conveys any information at all! As far as we can see from our present terrestrial perspective, messages from outer space are likely to involve the same type of redundant coding of information which has naturally evolved here on earth. Since redundancy implies non-random behavior throughout the text, this non-randomness must serve as our departure point in analysis of extraterrestrial communication.

Data of non-linguistic nature are also susceptible to analysis using the same approach as previously demonstrated. Suppose the following message were received.

Message #3

```
100000000000000000000000000000000000000001000011100
00000000100000100000010000010000000100010000000000
00000000000000000000000001000100001000000010000000
00000100000000000100010000010000001000000001000100
01000000011100000000000001000000000000010000000000
00000000000000000000000000000000000000000000000000
00001000001000000100000100001100010000000000110000
000000000000000000000000000011000011000011000000110
11011011010011000011001011001011001001001001001001
01010010010000110000110000110000110000110000000001
0000000000111110100000000000000000000000010000000000
01000001000000000001011011100100000000000001111101
00000000000000000000000000000100000000000000001000
10011100000000000101000000000000000101001001000011100
10101110010100000000000000000101001000010000000000010
01000000000000000010010000100000000000001111100000
00000000111110000001110101000001010100000000000010
10100000001000000000001000101000000000101000100000
00000000000010001001000010001001101100111011011010
00001000100010101010001000100000000000000000010001
0001001001000100010000001000000000000111000001111
0000011100000001111101000001010100000101000001000
00000010000000001000010000111000010000010000000100
000000000100000100010001000100000010000110001100001
000001000100010001000001000011000000001100001101
           10001101100000110011
```

Even to the naked eye the above string of bits seems highly non-random. The balance of bits is about 20% 1s and 80% 0s. More difficult to discover by eye alone is whether the 1s in the text distribute themselves randomly throughout the bit string. Using the Kappa test, as before, the coincidence rate should be about

$$0.20 \times 0.20 + 0.80 \times 0.80 = 0.68 \qquad (4)$$

assuming a random distribution of the 1 bits. The rates obtained in the test are as following for the first 50 shift intervals:

Shift	Coincidence rate	Shift	Coincidence rate
1*	0.71	26	0.67
2	0.69	27	0.68
3	0.69	28	0.68
4*	0.70	29	0.68
5	0.67	30	0.69
6*	0.73	31	0.68
7	0.68	32	0.66
8	0.68	33	0.67
9	0.67	34	0.68
10	0.69	35	0.66
11	0.69	36	0.66
12*	0.72	37	0.67
13	0.67	38	0.66
14	0.68	39	0.66
15	0.68	40*	0.71
16	0.69	41*	0.71
17	0.67	42*	0.71
18*	0.74	43	0.66
19	0.68	44	0.66
20	0.67	45	0.67
21	0.69	46	0.68
22	0.69	47	0.66
23	0.68	48	0.66
24*	0.70	49	0.67
25	0.69	50	0.65

Shifts which produce very high coincidence rates ($R \geqslant 0.70$) have been marked with an asterisk. There is a pronounced tendency for high rates to occur at shift intervals which are multiples of six, e.g. 6, 12, 18, 24, 42. This phenomenon appears to disappear after a shift of 42 has been reached. The clustering of high rates at the three consecutive intervals of 40, 41, 42 also calls attention to itself.

As previously, we now examine the coincidence pattern at positions yielding high coincidence rates in order to ascertain the cause of the phenomena. When we examine the coincidence pattern at a shift of 41, we find the following data where '+' indicates coincident bits and '−' indicates non-coincident bits.

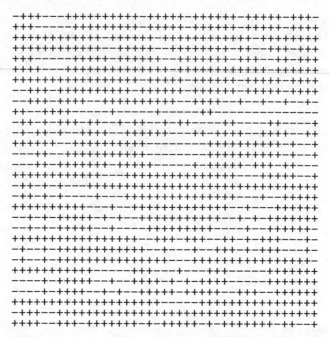

```
-+++---+++++++++++-+++++-++++++-+++++-+++-
+++-----+++++++++++-+++++-++++++-+++++-++++
+++++++++++-++++++-+++++++++++++-++++++++
++++++++++++--+++++--++++++++-++--+++-+++
+++-----+++++-++++++--+++++++-+++--++-+++
++++---+++++++++++++++-+++++++++++++-++++++
++++++++++++++++-+++++-++++++-+++++-++++
--+++-+++++++++++-++++++-+++++++-+++++-++++
--+++-++++--+++++++++++--++++--+--+--+--+-
++--++++-----+------+------++---------------
-++--+++-+++-+--+-+++-+-++----++-----++----+
-+-++-+--++++--+++++----++-+++++-+++++--++
++++++--+++++++++++--------+++++++++++-+--+
---++--++++++++++++--------+++++++++++-+--+
---++-+++++++-+++++------+-+++++-+++-++---
++++++++++++---+++++++++++++++++-+++++--+
-++-+-+---+++++++++++++++++++++++++++++++
-++-+-+----+---++++++++++++++++++---+--+++++
+++++++++---+--++++++++++++++---+---++++++
--+-+-+++++-+-+--+++++++++++---+-+-++++++
--+-+-+++++--+----++++++++++----+--++++++-
+++++++++++++++++++--+++-+++--++-+-+++---+-
-+---+-+++++++++++++-+----+--+++-+-+++---+-
-+---+-++++++++++++++--+++++++++++++++-+
++++++++++++-----+++----+---+++-----+++++++
----+-+++++--+--++++-+-+-++++-----+++++++
----+-+++++--+-+--+++++-+++++--+++--++++++
++++++++++++++++------+++++++++++++++++++
-+++--+++++++++++++++++++++++++++++++++++
++++--+++-++++++++-+-+++++-+-+++++++-++--+
```

The table has been written out in rows of 41 (the same as the shift interval). A great degree of symmetry is evident, with many coincidence patterns in parts of the table changing from row to row of the graph in an organized manner. This indicates that the data appears organized around columnar groupings corresponding to a row size of 41 bits. The other high coincidence rate shift intervals do not present such a high degree of columnar symmetry although some is present in each case.

We now write the original message in a matrix 41 columns wide. To make the result clearer, binary 1s have been replaced by '*' and 0s have been replaced by '.'.

This result is, of course, the well-known 1961 'bit picture' of Bernard Oliver (Callimahos 1966) which shows, among other things, a man and a woman each holding one hand of a small child while waving welcome to the lucky decipherers of the picture. Oliver's approach was to assume that the receiver of the text would note that it was composed of 1271 bits and that the number 1271 had only the factors 41 and 31. The factorization should immediately suggest the above columnar arrangement. We see that, even without resort to number theory, a little study along traditional cryptologic lines would have given the same result.

Oliver's bit picture idea has merit, real merit, since if we send an easily unscrambled pattern and send subsequent blocks of similar length, the receiver is quite likely to try the same bit transposition on following data.

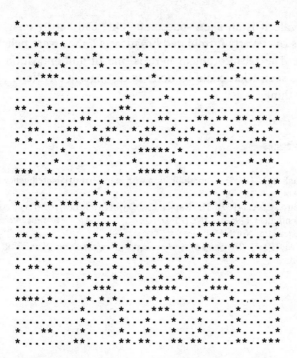

Thus, more complex pictorial data could be sent using the first solution as a 'cryptographic key' to following transmissions.

We have tried to demonstrate in this short sequel that the decipherment of any interstellar communication that might be received is likely to follow traditional cryptanalytic lines of attack. Further, even when the meaning of the text cannot be ascertained, we may learn a very great deal from its analysis, particularly if the sender has been generous with text material. This would enable a reply to be formulated which should, presumably, bring further aid from the other side.

It seems likely to the author that partial decryption of received messages may be easier than the traditional wisdom allows for, while interception may be more difficult than is now thought. The 'cosmic clue' we seek is the inherent non-randomness which we must have in any transmission so that a start may be made.

Another problem of some interest here is the possibility of forged messages. By this we mean to imply that messages may be received which purport to be of extraterrestrial origin but are not. One might attempt to cause political and economic disorder, or, perhaps the reverse, from such a move. For example, a satellite might be sent to the far reaches of the solar system and, once there, start transmitting back messages which seem to indicate it is an alien spacecraft. One can hardly imagine the consequences of cleverly formulated broadcasts if they were believed to be authentic. The basic problem here is one of authentication and it remains to be seen what could be done to distinguish any 'false alarms' from the real thing.

References

Callimahos, L. (1966). Communication with extraterrestrial intelligence. *IEEE Spectrum*, 3, 155–9.
Deavours, C. A. (1977). The Kappa test. *Cryptologia*, 1 (3), 60–7.
Freudenthal, H. (1960). *LINCOS: Design of a Language for Cosmic Intercourse*. Amsterdam: North-Holland.
Pope, M. (1975). *The Story of Archaeological Decipherment*. New York: Scribners.

Excerpts from *LINCOS: Design of a language for cosmic intercourse*[*]

HANS FREUDENTHAL

004. When dealing with artificial languages we must decide whether codes are to be counted among languages. First of all we shall explain the sense in which we will use the word 'code'. In the European languages 'code' may denote a collection of laws of jurisprudence, and in some of them 'code' is also the system of rules of etiquette. This circumstance may give rise to a terminology in which codes are distinguished from languages by their highly conventional character. In this terminology every artificial language would be a code.

I shall not accept this terminology. I shall distinguish codes from languages not by a genetic, but rather by a formal criterion. If the meaning of a message is to be kept secret, sender and receiver may use a language that is unknown to other people. Inventing and using brand-new languages is difficult. So they will try to parasitize on a common language. By a well-chosen system of superficial transformations they will change that common language into something like a language that can only be understood by people who are acquainted with the system of transformation rules. This collection of rules is a code in the proper sense (as a juridical code is a collection of juridical rules). If a code X is employed to transform plain English into a less understandable text, we say that this text is written 'in Code X', just as the original text was written 'in English'. So 'Code X', which was the name of a system of 'coding' rules, becomes the name of the linguistic matter which arises from transforming plain English according to the transformation rules of 'Code X'. I do not object to this ambiguous use of 'Code X'. The strong dependence of 'code language' on ordinary language is stressed by the fact that the same word is used for both the system of 'translation' rules and for the result of

[*] Amsterdam, North-Holland Publishing Company, 1960.
[Reprinted by permission.]

applying those rules. There is, however, still another word that means only the system of coding rules, viz. the word 'key'.

The relation between a natural language and a code is not the same as that between two languages. Translating from one language into another is quite different from coding and decoding. Coding and decoding can be done by formal substitution, whereas translating presupposes understanding. Of course this is again a gradual difference. The relation between spoken and written language is similar to that between a language and a code. Writing is like coding, and reading like decoding. Yet the rules that govern the relations between a spoken and a written language are much more complicated than the rules of any cryptography. The rules of writing and reading are not purely formal, as are those of coding and decoding. They can hardly be handled by people who do not understand the language in question, even in languages with a nearly phonetic spelling such as Italian and Dutch. Coding and decoding machines can be much simpler than machines for writing down spoken language and for reading written language. Furthermore, written language is much more independent of its spoken foster-mother than coded language is of its plain substratum. Spoken and written language may widely differ, and spoken language may undergo profound influences from written language. So it is not surprising that one speaks of spoken English and of written English as though they were really two languages. The relation between plain English and coded English is quite different from this. Calling coded English a language would not match our customary use of the word 'language'. Yet I would agree to coded English being called a *coded* language.

0 06. We rejected the denotations 'astronomical language', 'medical language', and so on for the idiom used by astronomers, physicians, and so on. We now resume this discussion because there is at least one exceptional case: the case of mathematics. Is there something that, in the frame of a reasonable terminology, may be called mathematical language?

A strong argument for an affirmative answer may be found in what happens if a book is translated from one language into another, let us say from English into French. If this book is a mathematical textbook or treatise, there will be parts that need not be conveyed into French, viz. the mathematical expressions and formulae. If we seek other examples of texts that are exempt from translation, the result will be meagre. First of all: the occurrences of Arabic figures. Yet Arabic figures are, to say least of it, so near to mathematical expressions that they can hardly be considered as a new instance. The proper names Shakespeare, Oxford, and

Portugal will reappear in the French translation in their original form, but "Pliny", "London" and "England" will be changed into 'Pline', 'Londres', and 'Angleterre'. Clearly it is not on principle that proper names are not translated. Or rather: the French translation of 'Shakespeare' is again 'Shakespeare', whereas the English and the French name of Pliny differ from each other. (Note that in a Russian translation all proper names would at least be transcribed, whereas mathematical expressions and formulae do not undergo any change.)

Another example: it is highly probable that a translator will leave quotations unchanged, especially if the language of such a quotation differs from that of the surrounding text. The sentence 'Before crossing the Rubicon Caesar said: Alea est iacta' will be translated 'Avant de traverser le Rubicon, César dit: Alea est iacta'. It is, however, not sure that a quotation of Hamlet's 'to be or not to be' would again read 'to be or not to be' in a French translation. (Perhaps adherents of formalist semantics might say that, as compared with proper names, quotations are not a new instance of texts not to be translated, because a quotation is a proper name, namely of the thing quoted. Yet I do not feel the need to use the term 'proper name' in a broader sense than usual.)

Even if mathematical expressions and formulae were proper names or quotations (I am sure that they are not), it would be a strange thing that they are never involved in the effective translation procedure. There are not many other examples of so strong an immunity. The best-known case is that of the formulae of chemistry. Physics is less striking. The resistant parts of a textbook on a subject of physics will be mathematical formulae, denotations of measures and units, and some reaction formulae like those of chemistry. The last instance of invariance with respect to translation I will mention are staves.

I shall restrict myself to invariant parts that are neither proper names nor quotations, and I shall ask the question: what is the reason for this invariance, e.g. why the mathematical formulae of an English textbook reappear unchanged in its French version. *A priori* two different answers are possible: the mathematical formulae are neither English nor French, or they are both. In the case of chemistry and music we may likewise choose between these two solutions.

We prefer the first answer: that the mathematical expressions and formulae belong to a language different from that of the surrounding context, because there are more reasons to adhere to this interpretation. The syntactical structure of 'mathematical language' differs enormously from that of all natural languages. The main points of departure are 'punctuation' and 'treatment of variables'.

... 0 11. A language that is to operate without regard to the meaning of its expressions, e.g. by people who do not understand it, must fulfil very high requirements. It must be fully formalized. Formal rules of constitution and transformation will determine the structure and handling of the linguistic expressions.

Formalization of language is a highly important problem. Coding as treated in 0 04 is an example of formal translation. Indeed, in such cases it is imperative that a code works without any regard to the meaning of the text to be coded or decoded.

This is a rather unproblematic example, because inside the coded text there is no need for operations, any more than for rules of operation. The coded language has a static character. The best-known example of an 'individual' really operating within a fully formalized language is the computing machine. Independently of any closer examination of the concept of meaning, we may assert that the machine does not understand the meaning of the linguistic expressions on which it is operating, in the sense in which we understand them. The machine works according to formal rules, whereas the human being who runs the machine knows the meaning of the input and output text. The subject matter on which the machine operates is language only from the point of view of the human being, just because it is he who attaches some meaning to it.

0 12. In view of the logistic tendency towards formalization it is not surprising that logisticians have paid less attention to language as a means of communication. In 0 10 I have explained why in the construction of a *characteristica universalis* we have to start with a concrete, sharply-defined, and rather narrow problem. After these explanations it will be clear why I think that this problem must be a problem of communication, and more precisely communication *ab ovo*. My purpose is to design a language that can be understood by a person not acquainted with any of our natural languages or even their syntactic structures. The messages communicated by means of this language will contain not only mathematics, but in principle the whole bulk of our knowledge. I shall assume that the receiver of these messages has understood their language if he is able to operate on it. He will be able to do so only if he has grasped the meaning of the expressions of this language, because it will be a moderately formalized language that cannot be handled on the ground of formal rules only.

The problem is not yet sufficiently outlined. Communicating with individuals who are not acquainted with our natural languages is a problem that is daily solved in the intercourse with babies and infants. We succeed in teaching them our language though we have started with a *tabula rasa* of lexicologic and syntactic knowledge. This linguistic education takes place in surroundings where things and situations are shown and named at the same time, by people who are acquainted with the language. Showing will mostly be unintentional; the act of naming will mostly include that of showing, even if the thing or the situation is not named for the purpose of being shown.

In order to narrow the problem, I propose to exclude or at least to restrict excessively the opportunities of showing. I shall use showing as little as possible as a means of explaining the meaning of linguistic expressions.

On the other hand I shall suppose that the person who is to receive my messages is human or at least humanlike as to his mental state and experiences. I should not know how to communicate with an individual who does not fulfil these requirements. Yet I shall not suppose that the receivers of my messages must be humans or humanlike in the sense of anatomy and physiology.

0 13. It was in this way that I arrived at the problem of designing a language for cosmic intercourse. Lincos, the name of this language, is an abbreviation of '*lingua cosmica*'. As a linguistic vehicle I propose to use radio signals of various duration and wave-length. These two dimensions will suffice. As compared with our acoustic vehicle, intensity as a third dimension is lacking. I rejected it because I think it less fit for radiographical use. But this is not a serious loss, because the wealth of information is much greater in the ether than in the acoustic medium, even if we dispose of the dimensions of duration and wave-length only.

I have not made a study of communication techniques, so I shall not dwell on technical details. When I designed Lincos, I had in mind using unmodulated waves. Perhaps modulated waves would be better. It would require only slight changes to adapt the program of messages to the use of modulated waves.

I have not examined which distances can actually be covered by radio messages. It is very probable that we can reach other planets of our solar system by means of radio signals, but I doubt whether with the powers now available we can communicate with inhabitants of other solar systems. Nevertheless I have taken into account the possibility that the receiver will belong to one of the neighbouring solar systems.

Of course I do not know whether there is any humanlike being on other celestial bodies and, even if there were millions of planets in the universe inhabited by humanlike beings it is possible that our nearest neighbour lives at a distance of a million light-years and, as a consequence, beyond our reach.

On the other hand it is not unthinkable that inhabitants of other planets have anticipated this project. A language for cosmic intercourse might already exist. Messages in that language might unceasingly travel through the universe, maybe on wave-lengths that are intercepted by the atmosphere of the earth and the ionosphere, but which could be received on a station outside. On such an outpost we could try to switch into the cosmic conversation. Long ago I thought cosmic radiation to be a linguistic phenomenon, but at the present state of our knowledge this seems to be very improbable. It is not easy to state *a priori* how to distinguish messages from purely physical phenomena. But should the case really arise, we shall know how to answer the question: we should try to understand the message.

This, I suppose, intelligent beings in the universe will do if they receive our messages. They will try to decipher them and to translate them into their own language. This task might be easier than that of terrestrial decipherers who have to discover the key of a code. Indeed our objective is just the opposite of that of the sender of coded information. We want to communicate with everybody who might receive our messages, whereas the sender of a coded message wishes to keep secret the information contained in his message. But in spite of our efforts even intelligent receivers might interpret our messages as physical phenomena or as music of the spheres.

0 14. It has been one of our presumptions that we cannot start cosmic intercourse by showing concrete things or images of concrete things. One might observe that television images are sent by radio and that we could try to do likewise. Television is founded on the principle that an area is decomposed into a set of spots and that these spots are arranged in a simply ordered (temporal) sequence according to a fixed rule. As long as we cannot tell the receiver the principle and the code of decomposition and arrangements, we cannot expect that he will understand television messages and transform them into images. Synthesizing television messages into images may be taught by means of analytic geometry. In our Lincos program, analytic geometry is a rather advanced stage. When this is reached, television, even three-dimensional, will be possible. But by then our linguistic proficiency will be great enough to dispense with television. I tend to believe that verbal description is superior to showing by television. I have not, however, paid sufficient attention to this problem as I am mainly interested in linguistic problems.

0 15. Throughout our exposition we will talk of Lincos as if it were speech (not writing). We shall say that Lincos phonemes (not letters) are radio-signals of varying wave-length and duration, and that Lincos words are constituted of phonemes (not of letters). If we aim at this constitution, we shall speak of phonetica (not of spelling).

This is no more than a convention, but it will prove very useful. We shall need to consider two versions of Lincos: Lincos proper, and a coded version. It will be convenient to use the term 'spoken Lincos' for Lincos proper and the term 'written Lincos' for coded Lincos. Moreover this choice of names is justified by the fact that the broadcast language will be 'spoken Lincos' whereas in printed expositions such as the present book 'written Lincos' will be used.

Written Lincos will not bear a one-to-one graphic–phonetic relation to spoken Lincos. On the contrary the relation will be global; every spoken Lincos word is reflected in written Lincos by a rather arbitrary graphical sign or group of graphical signs. This is a realistic procedure. A phonetic spelling of Lincos would be rather clumsy and illegible. Moreover for the time being there is no point in knowing how the different Lincos words will be pronounced (i.e. translated into groups of radio signals) when they are broadcast. The reader of the present book who meets with words of written Lincos such as 10, $+$, $\wedge$, Num, PAN, $♀$, and so on will understand that these graphical complexes are codings of spoken Lincos words. It is quite unimportant how these spoken Lincos words are constituted, and I have never concerned myself with this question. In any case the spoken Lincos word corresponding to Num need not be composed of phonetic groups which correspond to N, u, and m. In this book we shall stick to the convention that e.g. 'Num' is not the proper Lincos word but the written (or printed) image of it. But we shall not always stress this feature, and sometimes, in order to avoid clumsy language we shall even behave as if the Lincos word were 'Num'.

0 16. The notations in written Lincos are borrowed from mathematics (10, $+$), from logistics ($\wedge$), from other sciences ($♀$), and from Latin (e.g. Num is an abbreviation of Latin *numerus*). As a rule such abbreviations will be of three letters; we also use contractions of groups of written Lincos words, such as PAN (from Pau Ant Nnc). Finally there are Lincos words written in entirely arbitrary fashion, such as *Ha*, *Hb*, *Hc*, . . . which mean names of humans.

As pointed out before, the spoken Lincos correlates of the written Lincos words will be rather arbitrary. A certain restriction, however, should be made. Lincos will be more easily understood if it has a systematic phonetics. It would be pleasant if the various syntactic and semantic categories could be distinguished by certain striking phonetic characteristics. A receiver who has to interpret the Lincos messages, and who has made some progress in understanding them, will be grateful if he can see at first glance whether a word means a variable, a connective, a set, a human, a chemical element, and so on. Probably this will even help deciphering. Disconnected fragments of such a systematization exist in all natural languages. A noteworthy attempt to amalgamate these fragments has been made in Esperanto. There the syntactic and semantic categories emerging from Western languages have been accepted without much criticism and have been characterized by the traditional means of stems and inflections.

Establishing the principles of an adequate Lincos phonetics would be premature. No more than a rough systematization would be available at the present time. A few syntactic and semantic categories can be clearly distinguished at the moment, but their number is rather small. We have still to wait for a more extensive vocabulary of Lincos. Then we can resume this question.

017. In a more essential way the choice of Lincos phonetic means is restricted on account of the use of ideophonetic words. We have agreed to abstain as much as possible from showing, but we cannot entirely abstain from it. Our first messages will show numerals, as an introduction to mathematics. Such an ostensive numeral meaning the natural number n consists of n peeps with regular intervals; from the context the receiver will conclude that it aims at showing just the natural number n. Ostensive numerals will soon be superseded by algorithmic ones. An ostensive numeral is ideophonetic (as a hieroglyph both showing and meaning an eye is ideogrammatic). Other ideophonetic words are the so called time-signals introduced in the second chapter. These words have no simple correlates in other languages. One might even doubt whether they are really words, or rather signs, that mean nothing by themselves. More things will be shown in the Lincos broadcasting program, but then the means of showing are still less linguistic. With a common name such occurrences might be called noises. Some of them mean nothing but a noise, others mean unintelligible speaking, and still others mean events which are usually accompanied by a noise, such as in a radio play the noise of smashing a door means smashing a door.

The phonetic structure of an ideophonetic word is strongly determined by its meaning. It cannot be arbitrarily changed. This is even true of the less linguistic pieces of our program. Of course a noise meaning a noise may be replaced by another noise, but in any case the substitute ought to be again a noise.

There is still another kinid of truly linguistic phenomena which might be reckoned among ideophonetics: the means of punctuation. We have already had occasion to point out the importance of punctuation in mathematical and logistical language. Punctuation will be the principal means of showing Lincos syntactic structure. Lincos punctuation will be explained in 105_0. As punctuation marks we will use pauses, because pauses are self-explanatory to a high degree.

... $0\,21$. Lincos has to be taught to the receiver. Therefore our program to be broadcast cannot have the character of a newsreel. On the contrary, in the beginning we shall communicate facts which may be supposed to be known to the receiver. In this way we can try to facilitate the work of deciphering.

As our means of showing are heavily restricted, we cannot start with a concrete subject matter. Mathematics is the most abstract subject we know and at the same time a subject that may be supposed to be universally known to humanlike intelligent beings. So we have decided to start our program with mathematics. Natural numbers are introduced by ostensive numerals in contexts of elementary arithmetic. The first texts of this kind will be true formulae of the type $a + b = c, a = b, a > b, a < b$, where a, b, c are to be replaced by natural numbers.

Of course we cannot communicate all formulae of this type because of their infinite number. We shall communicate a number of such formulae that is large enough to justify the supposition that an intelligent receiver can conclude from these texts what the words (complexes of signs) $+, =, >, <$ mean. By a finite number of instances the relation between natural numbers described by $a + b = c$, or $a = b$, or $a > b$. or $a < b$ is not uniquely fixed from a formalist point of view. A machine that does not know any more about the relation $a + b = c$ but its truth value for natural a, b, c with $1 \leqq a, b, c \leqq 1000$ cannot solve similar problems with numbers a, b or $c > 1000$. Yet the receiver will be supposed not to be a machine but human-like. Children learn arithmetic by a finite number of instances, and a set of rather unformal rules. They succeed in generalizing the instances in a unique way. Arithmetic yields unique results though the definitions leave room for an infinite diversity. If I ask anybody to continue the sequence $1, 3, 5, 7, 9, 11$, I may be sure that he will do it with $13, 15, 17$ and so on. So I am sure that a humanlike being will conclude from a sufficiently large number of texts like $3 > 2, 7 > 1, 9 > 3$, that '$>$' has its usual meaning.

It is perhaps superfluous to insist on this point. A hundred years ago such assertions would have been self-evident. Since then we have got into the habit of giving formal definitions and proofs even for arithmetical and algebraical notions and formulae which could be defined and proved in a less formal way. It is not long ago that in arithmetic and algebra a common method of defining and proving was the method I have called elsewhere the quasi-general definition and the quasi-general proof. Euclid proves the existence of infinitely many prime numbers by showing that, three prime numbers being given, one can construct a fourth different from them. The proof is formulated in such a way that everybody understands immediately that it intends to show how n prime numbers being given, a $(n + 1)$-th prime number can be constructed. Euclid could not sufficiently formalize such a proof or even such an assertion because he lacked a functional symbol which meant numbering a general finite set, such as our system of numbering by indices, which we owe to a gradual development that started after Vieta and lasted nearly three hundred years.

Euclid's proof for the assertion A_n is a formal proof only for $n = 3$. For A_n it is 'quasi-general', i.e. every reader can be expected to be able to recognize the general pattern of the proof, and to generalize it for all natural numbers.

Up to the end of the 19th century it was commonly held that even geometrical proofs are quasi-general: a geometrical theorem is proved by means of one figure in such a way that the proof can be transferred to any other figure of the same kind.

Even now the quasi-general proof is still indispensable as a didactic tool, and the same is true of the quasi-general definition. In school mathematics, in sciences, and in daily life most definitions are quasi-general. One defines by means of instances, though in such a way that the other can generalize the definition as desired by the person who has given the definition. The method of quasi-general definitions (and even of quasi-general proofs) has been a fundamental principle in the construction and presentation of the Lincos vocabulary, in the chapter 'Mathematics' as well as in the later chapters. The reader will meet a good many examples of this method. However, we may remark that our definitions will never have the logical form of a definition. Things will not be defined, but described. We shall not say 'this is called a clock', but 'this is a clock'.

I will mention another didactic principle which has often been applied: If a set (a predicate, a relation) is to be introduced and named, we shall start not with a formal definition, but with a quasi-general definition, displaying a sufficiently large stock both of elements belonging to the set and of elements not belonging to the set (possessing the predicate, fulfilling the relation). So we shall not introduce the predicate 'integer' until other predicates can be named. We shall not introduce a word meaning action, until a large number of actions is known, and so on.

From the aforesaid it is clear that mathematical notions will not be introduced axiomatically. Of course when the Lincos vocabulary has developed far enough, one can add an axiomatic, and one can join this axiomatic to the preceding nonformalist mathematics by a semantic: we have renounced this, because it would not touch our main problem. In the chapter 'Mathematics' we shall restrict ourselves to the fundamental notions. We shall even drop notions of calculus which are needed in other chapters, because at present it cannot be our aim to write a textbook of calculus in Lincos.

0 22. This is closely connected to another feature of our project.

The definite program to be broadcast will consist of a large number of pieces called program texts. The collection of texts to be found in the present book is to be considered as an abstract from the definite program. It would be no use publishing something like the definite program in full. We have restricted ourselves to one example for every kind of program text which will occur. A kind such as $a + b = c$, $(a, b, c,$ natural numbers) is represented by the single example $4 + 2 = 6$, though actually the program will contain twenty or more of this kind. If variables are introduced as a new kind of word, every text containing variables will be repeated with other variables, so that a sufficiently large stock of words meaning variables and interpreted as variables by the receiver becomes available. Program texts in which certain humans display a certain behaviour are to be followed by similar texts in which a similar behaviour is displayed by other humans, so that the receiver can understand that this is a general behaviour, and so on.

Written program texts are enclosed in pairs of signs **#** which do not belong to the program text itself.

0 23. For the convenience of the reader the program texts have been collected in paragraphs, and the paragraphs in chapters. The first chapters bear the titles

I.Mathematics, II.Time, III.Behaviour, IV.Space, Motion, Mass.

Further chapters entitled 'Matter', 'Earth', 'Life', and a second chapter 'Behaviour' have been planned for the second volume of this project.

... **0 28.** Attempts to construct symbolic languages have often been linked to some philosophy, though the linking usually proved to be less close afterwards than the author believed. Dealing with language as a means of communication has been the only philosophy I have espoused in the present project. Though devised for a special aim, Lincos must be a general-purpose language in which all sayable things can be said. By formal linguistic means we cannot forbid any religious, ethical, philosophical, political or social faith or even any nonsense. So it is better to abstain from such attempts. Our method has been determined by linguistic considerations only. Our pragmatism has been that of an engineer, not of a philosopher. The use of both predicates and sets from the beginning does not prove any partiality in favor of nominalism or realism. The use of the word behaviour does not involve support for behaviorism. Introducing human bodies much later than humans themselves is not to be interpreted as a spiritualist attitude. Defining life as the possibility of perceiving is not the consequence of any form of sensualism. Lincos words for 'good', 'bad', 'allowed', 'forbidden' do not reflect any ethical theory.

Lincos is moderately formalized, but we do not object to fully-formalized language. Foolproof languages have an importance of their own. I hope Lincos will be wise-proof. Formalized languages appeal to the malevolent reader. The present book is addressed: *lectori benevolenti*.

Utrecht, December 23rd, 1957.

Chapter 1: **Mathematics**

1 00 0.   Pairs of signs # will enclose the printed image of a program text.

A (metatextual) 'and so on' after a text indicates that this text is an exemplary extract from the factual program. When carrying out the program we will replace this text by a large number of texts similar to the text we have printed. If the number of examples is large enough, we may expect that the receiver can generalize the quasi-general definition or proposition that is intended by the program text.

1 01 0. $\# > | < | = | + | - | \neq | \leqq | \geqq |$
        $. | .. | ... | .... |$
        $1 | 10 | 11 | 111 |$
        $a | b | c |$
        $\rightarrow | \vee | \wedge | ? | \leftrightarrow \#$

Loose Lincos words are presented, without any context, in order to stress their individuality. So it will be somewhat easier for the receiver to recognize them when they occur in a certain context.

The bold-faced strokes mean pauses.

1 01 1.   # . . . . . > . . . # and so on.

1 01 2.   # . . . < . . . . . # and so on.

1 01 3.   # . . . . = ... # and so on.

1 01 4.   # . . . . + . . = ... # and so on.

1 01 5.   # . . . . . . − . . = ... # and so on.

In these texts the Lincos phoneme that corresponds to the round dot is a short radio-signal (a peep). A Lincos word that consists of n successive phonemes of this kind, separated by short and equal intervals, is written as a group of n round dots. It both means and shows the *natural number n*. It is an *ideophonetic* word, which has the power of an image as well as that of a word. We also call it an *ostensive numeral*. The greater part of the Lincos vocabulary will be purely conventional; words may be permutated at pleasure. This is not true of ideophonetic words. Their essential features must not be changed.

The Lincos words written >, <, =, +, −, and so on designate connectives with the usual meaning. The receiver should guess their meaning from the context. Therefore each of the first program texts contains one unknown word only.

1 02 1.   # . = 1

       . . = 10

       . . . = 11

       . . . . = 100

       . . . . . = 101

       . . . . . . = 110

       . . . . = 1101 #

and so on.

Ostensive numerals are superseded by *algorithmic* ones, composed of *syllables* written 0 and 1 according to the rules of the *dyadic positional system*. For the convenience of the terrestrial reader we shall sometimes use the decimal code, but as a matter of fact such occurrences should be translated into the dyadic code.

1 02 2.   Texts as those of **1** 01 1–**1** 001 5 will be repeated, using algorithmic numerals instead of ostensive ones.

1 03 1.   # 111 = 110 + 1 = 101 + 10 = 100 + 11 = 11 + 100 =
        10 + 101 = 1 + 110 # and so on.

1 03 2.   # 111 + 11 > 11 + 101 > 1 + 100 = 101 # and so on.

In elementary arithmetic expressions such as '4 + 3' are often considered as problems, not as numbers; '4 + 3' is read as if it were 'add four
and three', '4 + 3 = 7' as if it were 'if I add four and three, I get seven'. Our
program texts will prevent such interpretations. The receiver will learn
the *relational symmetry* of the Lincos word written =, as well as the *'static'*
character of expressions as '111 + 11'.

1 04 1.   # 100 > 10
     100 + 11 > 10 + 11
     100 + 1101 > 10 + 1101
     100 + 1 > 10 + 1
     100 + 110 > 10 + 110
     100 + 11111 > 10 + 11111
     100 + a > 10 + a #

and so on.

By such texts we introduce Lincos words that mean *variables*. In Lincos
speech, words meaning variables will be distinguished from other words
by certain phonetic characteristics.

1 04 2.   # 100 + 111 = 111 + 100
     100 +   1   =   1 + 100
     100 + 1101 = 1101 + 100
     100 + 11 = 11 + 100
     100 + a = a + 100 #

and so on.

1 04 3.   # a + 1110 = 1110 + a
     a +   11 = 11 + a
     a + 11011 = 11011 + a
     a +   1   =   1 + a
     a + b = b = a #

and so on.

1 04 4.   Texts as those of 1 04 1–1 04 3 will be repeated using *other* words
that mean variables. A sufficiently large stock of such words should be
provided for. . . .

PART 6

Meaning and consequences of contact

We must at last consider what happens in the event of an extraterrestrial message being received. What will it mean, not in the sense of 'what does the message say,' but in the broader sense of what will be the significance – for mankind – of our learning that others exist in the universe? Should we reply, and if so, what should we say? If there is a possibility that we may somehow interact with the others – either in person or at a distance – then how should we treat them? Will they be persons in some sense? And if not, do they nevertheless have rights and interests that we must respect?

Because the possibility of contact is so remote in most people's minds, these topics are seldom considered seriously. But contact may occur at any moment (or, of course, never at all), and we might as well do some thinking about it in advance. Two of the authors in this section, Jan Narveson and Robert Nozick, have given these issues serious attention. But before we turn to them a prior question must be considered: How, in view of the time, expense, and resource-allocation that it will require, can a large-scale search for extraterrestrial intelligence be justified? This question is examined by Edward Regis Jr.

Regis, a science writer and Associate Professor of Philosophy at Howard University, Washington, DC, asks whether it is really true – as SETI proponents claim – that a search for extraterrestrial intelligence will repay huge dividends no matter how it turns out. Carl Sagan, for example, has said that the results of an extensive search 'whether positive or negative, would have profound implications for our view of our universe and ourselves.' For Sagan, finding out either that we are alone in the universe or that we have brothers across the light years would be discoveries of equally momentous consequence. Because of the deep and weighty implications which these discoveries would have, a large-scale search of the skies for signs of intelligent life elsewhere would in his view be worth every penny it costs.

Sagan's claims, together with similar allegations made by others, are critically examined by Regis, who finds them wanting in almost every case. Regis argues that for one thing the *non*existence of extraterrestrials is not something that can ever be proven. But if that is so, no consequences will follow from discovering that we are alone in the universe, for this is not something that we could ever in principle discover: there could always be intelligent aliens that we just have not found.

If, on the other hand, we do find intelligent life elsewhere, the implications of this will depend very heavily upon what we learn about those others and how similar they are to ourselves. If they are virtually identical to us, we will learn nothing at all from them, while if they are extremely different from us, then we will learn nothing from them in that case as well. Discovery of alien intelligence will have the profound and weighty consequences it is claimed to have, says Regis, only if the exceedingly improbable occurs, namely if contact is with an extra-terrestrial culture of just the right degree of similarity to and difference from ourselves. It is questionable, therefore, whether a large and extensive search is worth what it will probably cost.

Jan Narveson, Professor of Philosophy at the University of Waterloo, is author of *Morality and Utility* (Johns Hopkins, Baltimore, 1967), and editor of *Moral Issues* (Oxford University Press, 1983). Supposing that there is some contact between ourselves and aliens, he proposes five criteria to define our relationship to them in moral terms: the physical, sensory, mental, affective, and moral. Applying these standards to the three possibilities, that the aliens are superhuman, subhuman, or extremely different from ourselves, Narveson examines the moral lessons which contact with these different types of beings would teach us.

Robert Nozick, finally, projects a scenario describing the course of events following upon first contact. Nozick, Professor of Philosophy at Harvard University, is author of two classics in contemporary philosophy: *Anarchy, State and Utopia* (Basic Books, New York, 1974), winner of the National Book Award, and the recent and acclaimed *Philosophical Explanations* (Harvard University Press, 1981). In 1972 he wrote the short story that is reprinted here. Quite different in tenor and outcome from the tales told by SETI optimists, Nozick's account may equally well predict the long-awaited and oft-depicted first encounter between the others and ourselves.

SETI debunked

EDWARD REGIS JR

From the beginnings of SETI, its proponents have claimed that discovery of intelligent aliens will have momentous and profound consequences. Cocconi and Morrison in their original 1959 paper from which SETI science takes its origins say that 'few will deny the profound importance, practical and philosophical, which the detection of interstellar communications would have' (Cocconi & Morrison 1959). A year later Frank Drake said of the future discovery of extraterrestrial intelligence that 'the scientific and philosophical implications of such a discovery will be extremely great' (Drake 1960).

Although what these implications are is frequently not spelled out in detail, the usual assumption is that they will be uniformly beneficial to mankind. Bernard Oliver, a longtime SETI champion, predicts that 'Communication will yield a vast array of benefits' (Oliver 1976). Frank Drake and Carl Sagan, co-founders of the SETI movement, claim that contact with extraterrestrials 'would inevitably enrich mankind beyond measure' (Sagan & Drake 1975). Even SETI skeptics, who doubt there are extraterrestrials out there to be found, do not contest the point: Robert Rood and James Trefil, despite thinking that we are probably alone in the universe, concede nevertheless that 'the payoff of a positive search result would be very high' (Rood & Trefil 1981).

Recently SETI advocates have made the further claim that even an *unsuccessful* search would have profound consequences for humanity. In a 1982 petition, 73 scientists, including 7 Nobel laureates, urge 'the organization of a coordinated, worldwide, and systematic search for extraterrestrial intelligence.' They assert that the results of such a search ' – whether positive or negative – would have profound implications for our view of our universe and ourselves' (*Science* 1982).

But Carl Sagan makes the most surprising claim of all, that no matter how the search turns out – *whether or not* it proves the existence of extraterrestrials – major benefits will accrue to mankind. Failure to find extraterrestrials, he reasons, 'would have a profound integrating influence on the nations of the planet earth;' it would be 'a sobering influence on the quarrelsome nation-states' (Sagan 1982). Turning possible defeat into victory, Sagan claims that 'Even a failure is a success' (Sagan 1980).

SETI advocates are telling us, then, that the consequences of a major search – even if it finds nothing out there at all – will be so profound, the benefits to mankind so enriching, that we have every reason to go ahead and none for holding back. Since we can't lose, we may as well spend whatever it costs – upwards of $10 billion, according to some estimates (Project Cyclops 1973) – to make an exhaustive search of the skies.

These claims regarding the benefits of searching for alien life have gone largely unchallenged, indeed largely unexamined. This is unfortunate, for most of the claims are false.

I

If we are going to evaluate the thesis that contact with extraterrestrials – and even failure to establish contact – will have major and profound implications, then we have to know what such an implication is, at least as regards the question of extraterrestrial intelligence. Such an implication, I suggest, would be one which either changed our self-concept as humans in a fundamental way, or occasioned major changes in how we behaved toward one another. If, for example, a SETI search proved that the universe is run by Alpha Centaurians who control our lives from beginning to end, that discovery would have major and profound implications.

A distinction needs to be made, however, between, on the one hand, what success (or failure) of a major search would *entitle* or *justify* us to conclude, and, on the other hand, what people might, perhaps erroneously or without sufficient justification, think such a discovery means. For people have in the past attached wholly unwarranted meanings to certain events: an eclipse, for example, or a comet in the skies, have often but for no good reason been taken as portents of impending doom. In the same way, people might think failure or success of a SETI search implies something that it really doesn't. Although it would be interesting for us to speculate about the different meanings people might attach to the discovery of an extraterrestrial civilization, the focus in this essay is rather to say what such a discovery would entitle us to conclude, if anything.

We will begin with what would seem to be the most difficult case for the SETI proponent, namely to extract major and beneficial implications from a search that fails to find anything at all. Is the SETI enthusiast pulling rabbits out of hats, or is there a basis to his claims? To find out, let us assume that an extensive, lengthy search is made, but that no contact is ever established. What, if anything, would this entitle us to conclude?

The fact is that failure to find something out there would not entitle us to conclude very much. The problem is that even a major search's failure to detect an interstellar message would not by itself disprove the existence of extraterrestrials. For no matter how extensive and complete a search may be, the possibility would always remain that it was not complete enough, that maybe we had not searched the right places at the right times, or on the right frequencies, or with the right reception media. There are countless things that we might have missed, or even have failed to look for: gravity waves, tachyon messages, neutrino modulations, and so on and so forth. Since there is no way of knowing that any given search – or series of them – has exhausted all possibilities, there is always the residual chance that undetected aliens are still out there, that whole civilizations have escaped our notice. Indeed in the case of life on Mars this was just the type of argument made: to the news that the Viking landers failed to discover unambiguous evidence of Martian life, Carl Sagan's response was 'I haven't seen anything that makes me go negative on the idea of microbes on Mars. On the contrary, I'd have to say I stand about where I did before' (Cooper 1980).

Sagan's response is not atypical; in fact the SETI community has a more or less official policy about what absence of evidence for extraterrestrials means. In Freeman Dyson's words, 'the absence of messages does not imply the nonexistence of alien intelligences' (Dyson 1979). Or, as it is standardly put, 'Absence of evidence is not evidence of absence' (Project Cyclops 1973; Sagan & Newman 1983). But if this is so, then it is hard to see how the claim could be justified that the failure of even an extensive search would have any consequences or implications at all, much less that it would have beneficial ones. On the contrary it would seem rather that if we make a search and end up only with absence of evidence for ETI, then the only thing we would be entitled to conclude from this is that we have no evidence of their absence. In other words we are no better off – and no worse either – than we were when we started: we still don't know if they're out there.

We might, of course, express some surprise that our search had not turned up what we had expected to find. We might even have to revise our expectations somewhat. Alternatively, though, we could keep our expectations high and just revise our search modes. But from none of this would any significant conclusions follow.

Let us assume, however, that our search has been *so* exhaustive that the possibility of having missed someone out there is too remote to take seriously. Let us assume, in other words, that – somehow or other – we have been able to establish that there really are *no* intelligent extraterrestrials anywhere in the universe. Then what?

SETI advocates, as we have seen, attach major significance to such a result. Failure to find other civilizations, says Sagan, 'would tend to calibrate something of the rarity and preciousness of the human species' (Sagan 1983); it 'would underscore, as nothing else in human history has, the individual worth of every human being' (Sagan 1980). If it is really true that we are alone, the reasoning seems to be, then we are an exceedingly special species, and each one of us is much more worthy, precious, and valuable than we had heretofore supposed.

It is puzzling, however, to hear human specialness claimed to be a virtue when the pre-Copernican view that mankind is the center of things, and therefore special, is supposedly a 'deadly chauvinism' (Sagan 1983). For if it was wrong for the pre-Copernicans to have such distastefully ethnocentric, chauvinistic, and pridefully self-important notions about themselves (even when, in light of the available evidence, the pre-Copernicans' view that they were alone in the universe was well founded), then how is it somehow beneficial to return to those same notions once the possibility of extraterrestrials is ruled out? This would seem to be having our cake and eating it too.

And if we examine the claim that failure to find aliens 'would have a sobering influence on the quarrelsome nation-states,' we would have to wonder why the pre-Copernican conception of man's specialness failed to have this effect in years gone by. If, as seems to be true, the nations of those times were not more law-abiding and peaceful than modern states, then what reason do we have for imagining that a return to the old belief in humanity's specialness would have any such sobering influences today? This seems to be just wishful thinking.

In opposition to this, however, someone may say: 'But this is the 20th century. We're more enlightened today, and will be more sensitive to failures to find life elsewhere.' But the facts point otherwise. Return to the case of Mars. Prior to the Viking landers, many astronomers thought the chances were good that life would be found there. SETI theorists were of course in the forefront, and in 1962 Frank Drake had written 'it is almost certain that there is life on Mars' (Drake 1962). The failure to find

Martian life did not, however, have sobering effects upon the nations of earth. But, if the dashing of our hopes for life elsewhere in our very own solar system did not have the desired moderating influences, is there any reason to think that our failure to find life at even more remote locations in the galaxy would somehow produce those effects?

But here the SETI advocate may reply: even if the absence of aliens won't make a difference to people, nevertheless it *should* do so. They *ought* to act more cautiously, for the absence of intelligence elsewhere makes human life more precious here.

But why? Why should the absence of intelligent life in the farthest reaches of space and time give contemporary earthlings a reason to regard more highly the life that already exists here? It is true in general that the law of supply and demand operates with respect to goods: rarity increases their value. And the same thing holds, in a general way, of noneconomic phenomena: if the stars were visible on only one night in a hundred years, the sight would be more precious to all who beheld it. But is the same kind of thing true with regard to the human species as a whole, so that if there are no other like species out there, ours becomes more valuable or worthy of esteem, one that we ought to cherish all the more? This seems to be the SETI advocate's hope, but this doctrine has a fatal downside, for it makes the worth of a species – our own – hostage to what we may discover or fail to discover light years away from us. What is wrong with *this* is that, if *Homo sapiens* becomes more precious if extraterrestrial intelligence is absent, then in consistency it must become less so if extraterrestrial intelligence is present. On this doctrine, a universe teeming with intelligent beings would make human life mere tinsel.

I suggest that rather than to talk about the relative 'worth' of the human species, it may be better to speak of human rights and interests, for it is these that we want nation states – and individual humans – to respect. But if human beings have rights that we may not violate and interests that we may not harm, then our obligation to respect these things is as stringent if intelligent aliens exist as it is if they don't. The reason is that human rights are grounded, not in what might or might not exist on other worlds, but in earthly human beings themselves. Human beings are special not because they comprise a distinct species, and not because they are rare in the universe. Rather, humans are special because they are human.

It may be true that to find that we are alone in the world would be a discovery of great scientific magnitude: if nothing else it would present science with a very large fact to explain, one with which the special creationists would have a field day. But the *non*discovery of extra-terrestrials would have no further consequences, much less any beneficial ones. If we were to find that we are alone in the universe, we would be left much where we were before, and human life would continue more or less as it had been doing. And why not, since nothing would have changed?

II

So far we have considered what happens in the event nothing is found. But suppose that we *do* find something, that some day a message comes in, a genuine extraterrestrial intelligent message. It is tempting to think that this discovery would entrain behind it large philosophical consequences. But would it?

Here everything hinges upon the message's contents. If Deavours' argument elsewhere in this volume is correct, then it is possible to know that a signal is of intelligent origin without decoding the message it contains. But if we receive an undecipherable signal, then of course we will end up knowing nothing whatever about the sending culture. More optimistically, we may decipher a message only to learn that it is a representation of the prime number series or π to a thousand decimal places. At the furthest extreme we may tune in on signals encoding *Encyclopediæ Galacticæ*, or even more: 'The signal is . . . the song of people who have been alive, every single one of them, for a billion years. They are sending the information which will make this same immortality possible for all the creatures of earth' (Drake 1976). Although we cannot know in advance what an alien signal will say, it is probably as likely to be wholly unintelligible as it is that it will bring us the answer to life, the universe, and everything. Indeed the resistance to decryption of even some earthly texts – for example the Voynich Manuscript – does not inspire confidence that messages which are *truly* alien will ever be decipherable (Kahn 1973).

Suppose then that the worst happens, that we learn *solely* that there is (or was when the message was sent) another intelligent civilization in the universe, but nothing else at all. From this something significant does follow – the extension of the Copernican revolution in a dramatic and important way. We would know, after receiving the message, that not only is man not the center of things geographically (so to speak), he is not the center biologically either. He has competitors, perhaps even superiors. In an instant we would have plummeted from our special, pre-Copernican position to the status of 'just another civilization.' We would become common, in a way, typical, perhaps, or even pedestrian. We would have suffered a demotion. But surely *these* are not the promised beneficial consequences!

The *benefit* of finding another culture is that it would be 'a profoundly hopeful sign . . . . It means that someone has learned to live with high technology; that it is possible to survive technological adolescence' (Sagan 1980). It would be 'a powerful integrating influence on the nations of the planet earth. [It] would make the differences that divide us down here on earth increasingly more trivial.' (Sagan 1982).

Well, maybe. But there's equal room for doubt. For if *non*discovery of ETI means that humans become more precious and valuable, then it is hard to see why discovery of alien intelligence does not reverse matters. If aloneness in the universe puts us under a special obligation to survive, to husband earth carefully, then why doesn't our being one out of many other intelligent civilizations release us from that obligation? Why doesn't it mean that we can take chances, live dangerously, be careless with the earth if we wish? After all, if we should blow ourselves to pieces here on earth, there is another whole civilization out there waiting to take up where we left off. Made complacent by the knowledge that other intelligences exist, we can play fast and loose with the fate of the earth.

But of course no such conclusions follow at all. Just as nondiscovery of aliens does not magnify or reduce the scale of earthly problems, neither will discovery of aliens do so. For however large, serious or pressing our earthly difficulties were beforehand, they will be exactly as large, serious, and in need of solutions afterward.

We have prior experience to guide us here as well, the finding of new worlds right here on earth. The discovery of the Americas, for example, did not have anything like the effect upon Europeans that SETI advocates insist that discovery of ETI will have upon us. It did not make differences among Europeans more trivial, it did not serve as an integrating influence among them, it did not make them more tolerant and peace-loving. But if the discovery of a New World populated by other intelligent beings did not have the hoped-for beneficial effects back then, why should the finding of other worlds in space be expected to have any greater an effect now or in the future? It probably won't and indeed it shouldn't. Things are 'problems' on earth because they threaten or injure human well-being. If nuclear weapons, environmental degradation, and irrational hatreds threaten human life and welfare here, then we already have all the reason we need to minimize or eradicate these dangers. To find that there are intelligent beings elsewhere neither helps nor hinders in the least: the problems remain what they were beforehand.

But this is true only of the minimal case, where all we know is that another intelligent civilization exists (or at least existed at the time their message was sent), and nothing further. It would seem that additional implications, or at least *some* implications, would follow from our knowing more about the civilization in question.

Let us suppose then that we know a substantial amount, or perhaps even all there is to know, about the other culture. Then there are these alternatives: (1) We learn that their civilization is utterly alien to our own, so much so that we have virtually nothing common with them. (2) We learn that their culture is virtually identical to our own, so that we have virtually everything in common with them. (3) We learn that they are in some respects similar to ourselves, and in some respects different.

Let us suppose, first, that (1) is true, that – somehow – we learn enough about the sending culture to know that although they have intelligence in common with us, nevertheless in all other respects they are vastly different, probably more dissimilar to ourselves than we are from octopuses or bees. Perhaps the message comes from a world composed of a single organism (a *truly* 'organic' civilization) which, maybe in the way that Raup describes earlier, involuntarily emits modulated radio signals: a planetary whale singing for its own amusement. Or perhaps we detect an armada of von Neumann machines, and our incoming message is intelligent machinespeak, hash to us.

If anything like these things occur, then it is doubtful that we would be able to learn much from our alien friends. The reason is that *some* commonality is necessary in order for information in one civilization to be information in another. If the senders are as dissimilar from us as we are from ants, we have no reason to suppose that their transmissions will encode anything we can recognize, much less use. Far from expecting to get an ethical system from Them which will give us a new slant on things down here, They may not have any such thing as an ethics at all.

Would there be *any* implications, *any* consequences from such contact? It is hard to say whether our learning of such aliens would make us feel less or even more alone in the universe. For if we are to judge from many terrestrial examples of unfamiliar and exotic lifeforms, let alone from the fantastic creatures of horror stories and science fiction, *truly* alien creatures will be, at least to some degree, unsettling, perhaps even frightening. Should *these* be the extraterrestrials that we detect, not only will we not reap the promised beneficial consequences, we may even suffer some negative ones: we may inherit a sense of fear. With its strange and unnatural inhabitants, the universe may come to seem more alien than it ever did, and we might wish that we had never looked.

If an utterly foreign culture is a limiting case, then so is the opposite possibility, namely that the civilization we detect is exactly like our own, so much so that it is almost as if there is another earth up there in the skies.

No matter how unlikely this may be, let us suppose – for the sake of what lessons it might teach us – that there is a culture up there where the 'aliens' speak English, drive Fords and Chevrolets, and eat at McDonalds restaurants. *If* this astronomically unlikely possibility were nevertheless actualized, and if we received a message from them, well . . . we would hardly know whether to laugh or to cry. Just as in the case of a totally alien culture, there would be a distancing aspect to this discovery as well. I suspect that, after the shock of it all wore off, most people would find the existence of a duplicate earth to be an unbelievable disappointment.

One reason for the disappointment would be that – just as in the case of the utter aliens – there would be nothing for us to learn from them. These duplicate humans would have the same problems, and the same types of solutions to them, that we do (if they did not, they would not be the duplicates we have hypothesized). They would have the same ethical systems, the same disagreements, the same wars, the same everything. Their case would be so uninstructive to us that – again – we might as well not have made the search.

The third possibility is that we discover a species close enough to our own for there to be some applicability of their systems to ours, but yet different enough from ourselves for there to be something new and interesting in the messages they send. Rather than be identical with us, the others must be like us only in certain essential and relevant respects: they will have to be a society composed of discrete individuals having the same types of needs, drives, and desires as we have. They will have to be a culture in which resources are scarce and in demand, in which the interests of different individuals are sometimes in agreement, sometimes not, and so on and so forth. They will have to possess a language that we can understand, although that language need not be spoken, or even written.

A culture of this type will have the greatest chance of meeting the expectations of those who promise profound and beneficial consequences from contact: 'If the message contains valuable information, the consequences for our own civilization will be stunning – insights on alien science and technology, art, music, politics, philosophy and religion, and, most of all, a profound deprovincialization of the human condition. We will know what else is possible' (Sagan 1980).

But *will* the message contain valuable information? Aren't the prospects just as good that even their messages will be largely irrelevant or even unintelligible to us? Of our chances for tuning in on an *Encyclopedia Galactica*, Rood & Trefil say that 'unless the message was specifically tailored to a civilization just emerging into space, an extraterrestrial science book would be as incomprehensible to us as the wiring diagram of a radio would be to an aborigine' (Rood & Trefil 1981).

Recently, however, Sagan has suggested that it is only a message of this type – one intended specifically for 'emerging technical civilizations' – that we have any good chance of receiving, let alone understanding (C. Sagan, personal communication). But this suggestion is double-edged. For while it gives us some hope of being able to decipher their transmissions, it implicitly concedes that at least some alien messages may be directed toward civilizations incomparably more advanced than ours. Indeed, if the aliens' motive for communication is self-interested, then they are seeking to benefit from contact with societies more advanced than their own, just as SETI advocates on earth are. Communication with emerging technical civilizations, therefore, may be a low priority item for advanced extraterrestrials, in which case messages from them will be exceedingly rare.

Suppose, nevertheless, that we receive a message which is fully intelligible, and that it conveys a detailed account of the aliens' ethical or political systems. If these are the same as our own, then – yet again – we would have learned nothing new. If their ethics and/or politics are different from ours, that would be interesting, but it would also be problematical. For then we would have to decide what to do about them.

The only way to assess alien moral or political systems is to compare them to the systems of earth, and to project the consequences of adopting their doctrines here. Such an analysis would in the end lead us to embrace those aspects of their ethics that we thought would further human well-being, and to reject those that would not. But in doing this we would be using our own pre-contact standards. Call it provincial, reactionary, chauvinistic, or what you will, we'd be evaluating other theories by reference to our own. But to do this, to use our own accepted moral or political standards as criteria for the evaluation of other, alien, ideologies is hardly to get outside the earthbound perspective from which alien communications are supposed to liberate us.

Now someone may contend that this is just what we ought *not* to do: We should *not* evaluate their standards against our own, for to do this will entrap us in just the provincialism that we want to escape. But what is the alternative? Do we try the alien standards willy-nilly? Do we 'experiment' with them? What, for example, if the Others handle their population growth problems by practicing infanticide, or by killing off everyone above a certain age, or by random slaughters and executions? Do we courageously throw off perhaps chauvinistic moral objections to these practices and institute them for a while on earth?

As these questions illustrate, there is a way in which we are inevitably bound to the moral standards that we already have, and it is not altogether clear that this is a bad thing. But if this is so, then extraterrestrial theories may not provide us with solutions to our problems. They may not provide us even with interesting insights. They may not provide us with anything.

As for extraterrestrials alerting us to 'what else is possible,' we may know too much of this already. We already know that in addition to our own society's multifaceted, complex, often contradictory hodgepodge of moral and political doctrines, there are a multitude of other, competing systems that we regard as wrong. We also know that there have been a multitude more of these in the past, and that there are many more in imaginative literature, including, of course, science fiction. And while we cannot know in advance that every worthwhile ethical system has already been thought of, tried, and surpassed, the realization of the numbers that *have* been proposed and practiced by human beings past and present ought to make us skeptical of claims that the extraterrestrials will send deliverance from the skies. Further, given the extent to which earthlings have failed to learn from and apply the lessons of terrestrials in other cultures and other times, what reason is there to think that they would be at all moved by the moral or political notions of alien worlds long ago and far away?

So, to summarize, there is the initial problem of the applicability of extraterrestrial systems of thought and practice to our own: if their systems are not integrable with ours, they will be useless to us. But even if they should be applicable in principle, the problem arises whether we *ought* to adopt them. And while it is easy to say that a refusal to do so betrays close-mindedness, chauvinism and all the rest, the question must nonetheless be faced: *how else* could we decide what of theirs to reject, what to keep? In default of any better answer, we will have to say: by evaluating their standards in terms of our own.

What I have said of alien morals and politics is applicable analogously to possible extraterrestrial systems of art, music, religion, and so forth. In every case there is good reason to doubt that the incoming messages would provide the stunning insights that the SETI optimists have held out in promise. While it may be a matter of great interest to learn what beings on other worlds might believe, how they may act, what they may do, it is quite another matter to maintain that any of their practices or beliefs – provinicial to *their* species, time, place, and situation – will be of any benefit to us.

Matters are different when it comes to extraterrestrial science and technology. Here the problem of selectively accepting or rejecting alien doctrines in the light of our own does not arise, or at least not in the same sense. If, for example, the aliens provided us with the key to fusion power, if they showed us 'what else is possible' scientifically and technologically, these would be fundamentally important gifts.

This, however, presupposes that there are extraterrestrials out there who are doing our type of science (which Rescher, in this volume, has seen reason to doubt), that they have an interest in the same types of things that we have an interest in, have solved all the problems involved in a way that we could duplicate, that they wish to communicate this knowledge to others, and have tried to do so in a way that we can understand, and at a time which is congruent with our own. These are a lot of conditions to fulfil, and it may be asking too much to think that they are fulfilled very often, if ever. And while we have no reason to *deny* that these conditions are somewhere satisfied, it is optimistic to the point of sheer irresponsibility to claim that a SETI search ought to be conducted on the grounds that these kinds of returns are somehow guaranteed or even very much in the offing. Whereas SETI optimists uncritically take these benefits for granted, a more realistic assessment would have to be that a search based on the expectation of such returns is, in the words of George Gaylord Simpson, 'a gamble at the most adverse odds in history' (Simpson 1964).

As for the claim that contact with aliens will pay unprecedented dividends in our understanding of life and processes, Simpson's curmudgeonly response may be worth noting. 'The biological reward, if any, would be a little more knowledge of life. But we already have life, known, real, and present right here in ourselves and all around us. We can learn more from it than from any number of hypothetical Martian microbes' (*ibid.*, p. 775).

In summary, proponents of a major SETI effort contend that the time, expense, and work of the search are likely to be repaid in the form of profound pragmatic and/or philosophical implications, striking insights into art, music, and religion, and unprecedented advances in scientific understanding. These dividends are supposed to accrue, at least some of them, no matter whether or not we find extraterrestrials out there. We have argued, in opposition, that failure to find intelligent aliens will have no consequences for us, good or bad, and that if we do find them, we may end up unable to understand them or rejecting what they have to offer us anyway, on the grounds that humans cannot live, on earth, by the standards appropriate to other, quite different species elsewhere. So even after having been deprovincialized and opened up to new and different possibilities, even after having seen a vast new universe in front of us, we may in the end have to return to the earthly – and human – ways of living, acting, and knowing. Extraterrestrials might always be no good for us.

And yet . . . such arguments as I am making could be (and doubtless have been) given to oppose any new scientific investigation: 'Don't look into that, you won't find the answer, and if you do you won't like it!' But nothing of what I have said implies that we ought not to look. While I have tried to deflate what I perceive to be the uncritical and exaggerated claims of those who promise salvation from the stars, I have not replaced their assurances with others of my own about what we *are* going to get from contact. The cold fact is that neither proponents nor critics of SETI know what is up there nor what, if anything, They may be willing or able to tell us. Whether Their messages – if any – will in the end be useful is not something we can decide until we receive them.

I *do* claim, however, that SETI programs cannot in good faith be justified on the expectation of large-scale benefits or profound new perspectives. For this reason alone people ought not to be forced to support such programs through taxation – especially not to the tune of $10 billion. Fortunately, however, SETI searches do not have to be justified on grounds of legendary benefit, or financed by involuntary taxation, in order for it to be worth our while to continue searches already in progress. For there will be one undeniable benefit if one day we have a positive result: it will satisfy our curiosity to know what else is out there. While this will be far less than what SETI enthusiasts promise, it may well be all the motive we can ever have to undertake the search they wish to make.

For their comments on a prior draft of this essay, I am indebted to Pamela Regis, Charles Griswold, Lewis White Beck, and Carl Sagan.

References

Cocconi, G. & Morrison, P. (1959). Searching for interstellar communications. *Nature*, **184**, 844–6. (Reprinted in Goldsmith, *op cit.*, 1980.)

Cooper, Jr, H. S. F. (1980). *The Search for Life on Mars*. New York: Holt, Rinehart and Winston.

Drake, F. D. (1960). How can we detect radio transmissions from distant planetary systems? *Sky and Telescope*, **19**, 140–3. (Reprinted in Goldsmith, *op. cit.*, 1980.)

– (1962). *Intelligent Life in Space*. New York: Macmillan.

– (1976). On hands and knees in search of Elysium. *Technology Review*, **78**, 22–9.

Dyson, F. (1979). *Disturbing the Universe*. New York: Harper & Row.

Goldsmith, D. (1980). *The Quest for Extraterrestrial Life*. Mill Valley, California: University Science Books.

Kahn, D. (1973). *The Codebreakers*. New York: New American Library.
Oliver, B. (1976). Why search? In Goldsmith, 1980.
Project Cyclops (1973). Moffett Field, California: NASA CR 114445.
Rood, R. & Trefil, J. (1981). *Are We Alone?* New York: Scribners.
Sagan, C. (1980). *Cosmos*. New York: Random House.
– (1982). The search for who we are. *Discover*, 3, 31–3.
– (1983). We are nothing special. *Discover*, 4, 30–6.
Sagan, C. & Drake, F. (1975). The search for extraterrestrial intelligence.
    Scientific American, 232, 80–9.
Sagan, C. & Newman, W. (1983). The solipsist approach to extraterrestrial
    intelligence. *Quarterly Journal of the Royal Astronomical Society*, 24, 113–121.
    (Reprinted in this volume.)
Science (1982). An international petition. 218, 426.
Simpson, G. G. (1964). The nonprevalence of humanoids. *Science*, 143, 769–75.
    (Reprinted in Goldsmith, *op. cit.*, 1980.)

Martians and morals:
How to treat an alien

JAN NARVESON

Certainly one of the reasons for being fascinated by the possibility of space exploration lies in what might be found out there. The prospect of encountering some sort of intelligent life on other planets, for example, has intrigued people for a long time. In fact, the latest word is that the prospects for such an encounter are dim. Recently obtained information about the environments available on the other planets of our own solar system rules out any reasonable chance of life beyond the most rudimentary level. The probability that there is intelligent life elsewhere in the universe, on the other hand, would seem to be overwhelming. For there are, after all, several million galaxies in good working condition, and each of those contains, on average, several billion stars. Against this, however, we must set the impossibility of our ever encountering any of these beings, or vice versa. The nearest star is about four light years distant. This means that a spacecraft travelling at a really good clip – say, 100000 miles per hour – would take about 7000 *years* to get there. It also means that communication with people inhabiting planets of that star, in the highly unlikely event that it happens to have planets suitable for the evolution of human-level life, would be extremely inefficient, since a simple exchange of messages would take eight years to complete! Relative to current information, therefore, there would seem to be no reason to expect any of the encounters envisaged in *Star Wars*, for example. In consequence, the subject of this essay is tantamount to fiction, its suppositions purest fantasy.

Nevertheless, we might be able to derive some instruction from suitably directed fantasy. Projecting ourselves into these almost purely imaginary situations may help to bring out things that we need to know regarding the moral encounters we regularly have in our lives. I refer, in particular, to the general question of differences between ourselves and others: which of these are *morally relevant*, and which not? Questions of this kind arise constantly. Are fetuses sufficiently like ourselves to be accorded the same range of rights as adults? What about comatose humans? How about animals? Such questions are often asked as though we know how we ought to treat fellow normal adults: and this is a fairly reasonable assumption. But reflection on the unusual case may make us realize that our standard way of assessing our moral relations to other people is not as encompassing as we would like it to be. Might this happen in the exploration we are about to make? Let us see.

'Morally relevant'

The question of moral relevance is the question of what makes which moral differences. Suppose we know how to treat A and know that B is just like A except that F-ness distinguishes A from B: B has F and A does not. Does it follow that we are to treat B differently, given that we knew how we should treat A? If it does, then 'F' is a morally relevant predicate. That is the general idea, anyway. But the general idea is still not perfectly clear. A great deal of weight has been put, especially in recent moral philosophy, on an apparently uncontroversial 'Formal Principle,' to the effect that 'It cannot be right to treat A in one way and B in another unless there is some difference between A and B other than the fact that A is A and B is B'.[1] Yet if there is a glass of water on the table before him, and A knows that B is just as thirsty as A, is it not nevertheless right that A procure the glass for A rather than B, even if it would be just as easy for him to give it to B?

The point of this example is that the term 'right' is ambiguous. Let us distinguish two senses of 'right', which we may call the 'strong' and the 'weak' senses:

> Strong: 'x is right' = 'it would be wrong not to do x'
> Weak: 'x is right' = 'it would not be wrong to do x (but would also not be wrong to do at least some alternatives to x)'

Thus it is right in the Weak sense to wear brown shoes, but certainly not in the Strong sense (in any normal circumstances, anyway). But it is right in the Strong sense to tell the truth (in normal circumstances, anyway), to have regard for the safety of persons around one (same caveat), and so on. We can correlate this distinction with other items in the moral vocabulary: if x is *Strongly* right, then we will say that you 'ought' to do it, or 'must,' or that you are 'required,' that it is a 'duty' to do it; whereas if it is only *Weakly* right, then we will say that you 'may' do it, that it is 'OK' to do it, or that it is 'permissible' to do it, but not that you 'must,' etc.

It is evident that our formal principle must be employing the term 'right' in its Strong, rather than its Weak sense. There are plenty of differences between one person and another which I am perfectly entitled to take into account in my normal behavior. I shall go far out of my way to hear *A* perform Beethoven's '*Hammerklavier*' *Sonata*, but scarcely out of my way to hear *B* do so, even though the difference here – that *A* is a superb pianist, *B* an indifferent one – is not one that I, at least, would associate with a difference of moral duty. (An 'aesthetic duty,' perhaps; but that is another matter.)

But this raises a point about the term 'morally *irrelevant*'. Some will say that it is morally wrong to treat people differently on the basis of morally irrelevant differences; but this can only obviously be true of what we may now call 'Strongly irrelevant differences' – differences which one has a moral duty *not* to take into account. Yet most differences which tend to be labelled morally irrelevant are, I think, only Weakly so: some obvious examples are differences of race, sex, appearance, talent, and so forth. Yet there are plenty of cases in which one may take these into account in one's behavior. I know people who are what might be called 'reverse racists:' they seek out the company of persons of other races, inviting them to dinner, more readily making friends with them, and so forth. Is their behavior wrong? Does morality require them instead to randomize on that particular variable? I do not see that it does. What is obvious is only this: that in formulating the *fundamental* principles of morality, morally irrelevant differences may not be cited in an essential way. This is obvious because it is, after all, true by definition: a difference is morally irrelevant if it does not make a difference in one's duties toward the person or things in question (in the respects in question).

Turning to extraterrestrials, then, we must ask: what sorts of differences between them and ourselves would make for differences in the ways we ought to behave toward them? Here, everything hinges on the amount and type of difference in question. For, in the special case where the extraterrestrials (henceforth ETs) are virtually identical to us in their biological and genetic make-up, and where their resources and ecological situations are analogous to our own, presumably we ought to treat them more or less as we should treat other humans.

It is more likely, however, that ETs would be very different from us, not only in their situations and resources but also in their persons. It would be rash to try to lay down limits on the degree of variation from the human case that might be possible for ETs. But our concern here is with morality, rather than just any or all possible contexts in which we might be sensitive to such variations, and so our question is which possible differences of person would affect our possible moral relations with aliens. Some such differences would, surely, make any kind of recognizable moral relations impossible. Others, we may suppose, would alter the *content* of our relations without appreciably altering the *form*. The most interesting cases, though, would be of beings that are recognizable as persons, but are persons of such an extraordinary nature as to alter the form of morality, yet leaving it recognizable as morality. This latter possibility, I take it, would have been denied by Immanuel Kant, for instance, and that view must certainly be taken seriously. Indeed, I think, in the end, that he is right.

In using the terms 'content' and 'form' above, I mean to indicate a sort of difference that readers may find opaque. And rightly so: it is also opaque to philosophers, at least if we can judge by the amount of dispute about this sort of thing that goes on among them. Be that as it may, I will try to give a general idea of the sort of distinction intended. Consider the following moral precepts:

   (i) All persons ought to refrain from killing other innocent persons.

   (ii) All persons ought to refrain from shooting other innocent persons.

Shooting is a specific way of killing people. Relative to shooting, killing is (more nearly) formal, shooting more nearly a matter of 'content.' It would be easy to come up with a (plausible) more formal principle than (i): e.g. that all persons ought to refrain from *harming* other innocent persons. But shooting, killing, and harming are all 'material;' they are matters of 'content' rather than 'form.'

Some ethical principles have been stated so abstractly that no immediate implications concerning which particular things one is being told to do or refrain from can be drawn. Thus consider:

   (iii) Do unto others only as you would have them do unto you.

What (iii) implies about conduct depends on how the agent 'would have' others treat him or her, which will vary from person to person. Nevertheless some have supposed that (iii) is perhaps the supreme principle of ethics. Kant, in particular, has advocated another version of principle (iii), namely

   (iv) Act only on that maxim that you can will to be a universal law.
        (Cf. Kant (1785).)

Principle (iv) is known as the Categorical Imperative (or more precisely, the Supreme Categorical Imperative). Happily, it is not our business here to decide on the merits of Kant's proposal, which are still being hotly debated. But Kant's principle well illustrates the category of 'formal' principles, in one relevant sense: a very abstract, high-level principle which would require further information to apply it to any particular case.

Another relevant sense of the term 'formal' here is that a formal principle is one which specifies, or formulates, the 'form' of a subject, the form being what is essential to it, what makes it the subject it is. In the case of morality, there are, as we know, many 'codes' of morals, differing in detail or even substantially from one society or even from one person to another. But what is it which makes all of these things moralities, moral codes rather than something else? Whatever it is, that would be the 'form' of morality in this further sense of the term. Kant and many others supposed that there is one, or at most a very small number of, principles that anything purporting to be a moral code would have to include as a first principle. The Categorical Imperative, on Kant's view, was the supreme formal principle of morality in that sense: any plausible candidate for a 'morality' would have to contain principles that could be plausibly supposed to follow from, or be specific applications of, the Categorical Imperative. The formal principles of morality are those that give us our sense of right and wrong; they are what make the subject recognizable, so to speak. (They do this, even if we do not know what they are: they are to be the 'underlying' principles, for which we might have to do a lot of intellectual digging, as in the work of Kant and others.)

When I ask how the differences between ourselves and possible ETs affect the 'form' of morality, I am asking whether these differences might be so profound as to alter our entire sense of how other beings ought to be treated. Might ETs differ from us so much that we no longer think that we ought to do unto them as we would have them do unto us? So much that the whole idea of 'acting only on those maxims we could will to be universal laws' seems laughably out of place?

My answer to these questions, by the way, is No (as will be seen later). Some differences between ourselves and other creatures would make a vast difference to how we ought to behave toward them. In fact, many differences between us and other creatures could be such that it would be literally impossible to do acts of certain kinds to them at all: you cannot break the arm of a creature that has no arms; and there is a sense in which you cannot lie to a creature that can read your mind. On the other hand, we cannot tamper with the mind-waves of our fellow humans, because they do not (so far as I know!) emit mind-waves. But maybe Alpha Centaurians do. In examining the various possibilities that the thought of ETs raises, at least in my mind, I hope also to add something to our grasp of morality in the mundane world we currently occupy.

We need a classification of possible differences for this purpose: in what respects might alien 'beings' differ from us? Here is an attempt at a general classification:

(1) *Physical:* Anatomical differences, differences of the stuff of which the being is composed (might there be inorganic beings of interest?), in its diet or fuel, etc.

(2) *Sensory:* Perhaps the being comes equipped with sensors that react to very different portions of the gamut of physical fields – light, magnetism, etc.

(3) *Information-processing:* Here there are two sorts of differences to be concerned about: (*a*) the ET might be much better or worse at recognizably similar operations (e.g. logic); (*b*) but perhaps it would perform operations of a type unintelligible to us, as Rescher (this volume) has suggested.

(4) *Affective:* An ET might differ in its emotional capacities, e.g. by simply not having any at all, or by being affected in very different ways by very different stimuli. Some ETs might strike us as hopelessly neurotic, others as uncannily, even incredibly, stable.

(5) *Values:* The life goals of aliens might be so different from ours as to be doubtfully recognizable. How much can they differ from us, and still be perceived by us as being 'life goals'?

The differences that these various classes might have are, as one would imagine, quite different. Differences in categories (2) and (3), for example (and conceivably in (1) as well) could render communication with the beings in question exceedingly difficult or impossible, and this would surely affect their *eligibility* for moral relations with us. Other differences, especially in respect of categories (4) and (5) (and to a lesser extent, perhaps, (1)), would alter drastically the *content* of any moral relations we might have. Whether any of them promise an interesting alteration in the *form* of morality, yet leaving our transactions with them recognizably moral, is perhaps the most interesting question, and I shall return to it.

A sort of difference not mentioned on the foregoing list has to do with what might be called Personnel Organization. We are accustomed to thinking in terms of 'individuals,' organisms with definite, unified control centers (i.e. minds) whose decision-making apparatus extends to the physical limits of that organism and whose relations with other organisms of its own or any other kind are 'external' in the way that gives rise to familiar philosophical problems such as the problem of Other Minds. But such developments as computer networking suggest the possibility of organisms whose sociality is of a different order from this. We can imagine that some types of organisms are somehow interconnected in such a way that the applicability of the notion of 'individual' to them would be extremely problematic or even hopeless. Imagine issuing the sort of mental 'command' that one normally does in reaching for one's coffee cup, only to find that the hand of some other person responded instead. Or that one kept finding readouts on one's own mental TV screen which were the thoughts of other people. I have described both of these possibilities in our familiar conceptual scheme in which we can identify commands and responses as being those of distinguishable individuals; but if we were to multiply instances of the sort of phenomena mentioned to the point that they were utterly typical of our cognitive/motor/affective lives, then we would have a situation in which a fairly drastic realignment of such concepts would seem to be called for. And I am ready to concede that this might have rather drastic effects on the application of ethical notions to such entities as well. So drastic, indeed, that I am at a loss what to say about them: how do we talk about right and wrong in the relationships of beings who are not distinct individuals and thus cannot easily, or perhaps at all, formulate the notion of 'duties to others'?[2]

In fact, idealist philosophers have been telling us for quite some time now that ordinary humans are already like that, and that 'our' notion of 'an individual' does not apply even now to the very people we know. It seems not much use replying to these philosophers that, if their view were correct, then possibilities of the kind obscurely envisaged in the preceding paragraph ought not to strike us as bizarre as they do; and the fact that we are all more or less identical with each other seems not to prevent gangsters, terrorists, and soldiers from killing what they (mistakenly?) believe to be 'other' people. But it is of interest that we might, with the example of computer networks before us, be able to go some way toward imagining a set of minded entities that provided an intuitively more plausible model of the idealist's theory. The question also arises, however, how the possibility that some set of beings with whom we came in contact were non-individual (in some such sense) would affect *our* moral relations with them (or rather, I should say, with 'them/it'). And off-hand, it is difficult to see why we should not deal with them as a highly unified nation – which, in fact, is how government ministers of states deal with other states even now. Our imputation of unity to most states at present is metaphysically fraudulent, for the most part, but if we came across a set of entities with minds/mind, organized in the collective way envisaged, would the imputation in question be any the less fraudulent? I think not; but at any rate, we shall have to leave this tantalizing possibility for another time.

Meanwhile, in order to impose a manageable order on our inquiry, I now wish to suggest that we consider differences of the kinds mentioned under three headings. Namely, we will survey each kind of difference as it would reveal the beings in question, in the respects in question, to be (1) Superhuman, (2) Subhuman, or (3) 'Other' – that is, not comparable (enough) with the human situation to enable the first two kinds of comparison to be drawn. We start with the first.

Superhumans

In all of the classes of difference, but more especially in the first, second and third, it would be natural to describe some possible ETs as 'superhuman'. In popular science fiction, I suppose, these are the most frequently encountered sorts of beings. In physical respects, for instance, we can imagine creatures that, by human standards, could perform miracles of object-manipulation. We imagine spider-like creatures of roughly our size that could jump over very high buildings, crush any human with a mere pinch of its fingerlike organs, etc. We should also consider the possibility that the beings might be essentially *invulnerable* to anything we might try to do to them – though considering the destructive capabilities of thermonuclear weapons, it is a bit difficult to believe that any such creatures could exist. In their sensory capacities, there is the possibility that they have X-ray vision, or that they directly detect all sorts of signals we do not even know exist or that we know about but need elaborate apparatus to detect. And of course nowadays we can readily endow such hypothetical beings with incredible information-processing abilities. In describing these various attributes as 'superhuman,' what of possible *moral* significance might emerge?

The 'superhuman' side of the ETs in question does not – as will no doubt have struck the thoughtful reader (or TV viewer) – consist in their being very, very, extraordinarily human. It consists, rather, in their being extraordinarily powerful or capable in ways understandable to us. Whether these properties are of moral significance will surely depend, however, on their psychologies. What are they likely to do with these powers? It is logically possible that beings with superhuman powers should also be enormously helpful to us, even kind and generous. That is not what is usually supposed; but if it were so, then surely the implication for humans would be that we should refrain from rendering the creatures unlikely to use their great powers on our behalf. If the creatures were also invulnerable, of course, this would be otiose; but suppose we credit them with an Achilles' heel – a likely candidate being vulnerability to nuclear bombing – then the principle that one is not to do anything that crosses the ETs would surely be strongly recommended.

What sort of things might they be able to do for us? The possibilities are plainly limitless: one can imagine beings that are super-sweepers, just crazy about keeping floors spotless and getting them into that condition with incredible efficiency; or super-builders, ready to construct a sky-scraper out of inexhaustible handy materials in hours and any time one liked; or . . . . If there were not very many of such creatures, one can readily imagine the usual range of human problems arising about the allocation of their extremely available services, probably leading to wars and divisions of capitalists and socialists, and so on. All of these are, of course, *human* problems, as I say, and it is difficult to see that anything of very fundamental moral significance emerges. If the creatures in question were able to perform at a level that made their capability of satisfying human demands essentially unlimited, so that no problems of allocation of scarce resources arose, we would then be in Hume's Garden of Eden situation, which would make the usual range of quarrels with respect to the goods thus made available utterly pointless and, presumably, would thus even lead to their cessation.[3] In such a case, a quite different kind of problem might arise: if life is too easy, will humans be able to prosper and flourish? One readily imagines that an ethic of self-denial would arise, calling upon us to refrain, in our own long-run interest, from indulging ourselves to excess in the ETs' services. This is a recognizable situation: when one has an automobile, and lives a fair walk from one's place of work, there is always the temptation to take the car, although the exercise from walking might be just what one needs. And so on. Again, this does not seem to be of fundamental moral interest.

But we may suppose it more likely that ETs that were recognizably persons with super powers in some respects would not derive their highest delight in life from serving any human who wished it with total and slavish devotion. One possibility, certainly, is that the ETs would be just the reverse of that, determined instead to enslave us, or perhaps to raise us, like cattle, for their dinner. Let us contemplate this unwelcome possibility. There are, it would seem, two important versions to distinguish. One is the case where the beings in question are invulnerable to anything we can do to them; the other is the case where they are not. (This makes it sound much too simple, of course. Clearly there would be an endless array of distinguishable possibilities in each case.) Let us consider each in turn.

My term 'vulnerable' is not entirely satisfactory for the purpose

intended here, for it tends to be used in connection with liability to harm
or damage. But we should distinguish two subcases: (1) the standard use
– liability to be killed, injured, immobilized, etc.; and (2) the less usual
use that consists in our being able to withhold desired services or goods.
The difference is significant, for one can easily imagine that one sort of
being might be utterly unable to inflict damage on another – the others are
bulletproof, etc. – and yet that the others have a great interest in things
which we can uniquely provide for them. In a much lesser degree,
something of the kind obtains between human males and females: men
are much more capable of inflicting physical damage, mortal and otherwise, on women than vice versa, but women can provide things that few
men would want to do without if they could help it. And thus, despite the
on-going 'battle of the sexes,' men and women do manage to get along
frequently, even typically. It is not difficult to imagine a relation between
humans and another intelligent species being similar to the relation
between women and men. And in that case, there would be hope of
peaceable and mutually profitable relations with the beings in question.

But what if, instead, humans are to ETs as ants are to humans? Suppose
a race of ETs so virtuosically superhuman that mere humans stand to
them as completely insignificant. There would seem to be two questions:
first, whether this is possible, and, secondly, what follows if it is? The
latter, I think, is the easier to answer. Some philosophers, I believe,
would be inclined to argue that, since the superhumans would owe us the
duties associated with the idea of treating persons as ends in themselves,
we should therefore take a much kinder view of ants than we are
accustomed to. But the argument can surely go in the other direction with
more plausibility, if not more comfort to our egos: we could conclude that
the superbeings would simply have no duties to us – that at best we might
be regarded as objects of scientific curiosity to them, just as the ants (apart
from their nuisance value) are to us. From this we may derive such
comfort as we can, which is to say frankly, not much.

Whether it is possible for superbeings to compare with us in that way is
an interesting question. It might be argued[4] that there is a difference in
kind between us and ants, not merely one of degree, and that *that*
difference makes all the difference. In particular, we are conscious –
indeed self-conscious – beings, with all that that entails; whereas, we
suppose, ants simply have no interior life, no consciousness at all. We
'suppose' this, but do we actually know it? That is difficult to say.
Consider the human fetus. Is it conscious at all? Electroencephalogram
tests and suchlike are alleged by some to indicate that they are not, indeed
even to prove that they are not. And certainly I side with the negative

here, but not because of my familiarity with fancy scientific equipment. My difficulty is simply that I cannot recall ever having any sort of recollection of what things were like when I was a fetus, nor can anyone else whose opinion I have troubled to obtain. If there is any fetal consciousness, it is evidently not substantial enough to leave any phenomenal memory traces, and that is surely not enough to credit it with consciousness. In the case of ants, though, the problem is plainly not that. Despite the sophistication of their behavior in some curious respects, we just do not find it possible to draw anything but a blank when we try to imagine our way into the mind of the ant. But this, unlike my proposed test for fetal consciousness, is surely inadequate: the fact that we cannot get anywhere interesting in an effort to envisage the consciousness of the ant may just show a limitation in us rather than the ant.

Or it might conceivably show that we are quite unable to sympathize with these (individually) insignificant beings. We have, after all, a pretty good idea of what ants can and cannot do, whether or not we can manage to suppose that they are conscious beings. And suppose they were. Suppose ants do intelligently figure out how to build ant hills, that when they have a parade in some direction, in which all the others queue up efficiently behind the leader, it is because they *want* to follow the leader and appreciate the vital importance to the general welfare that they do so. Are we now supposed to be impressed? Impressed to the point where we will perforce grant rights to the ants? I think not, actually. Instead of being terribly impressed with the fact that they were after all conscious, intelligent beings, why should we not instead be terribly unimpressed with their philosophies of life? Does a being which would intentionally devote his entire life to the things to which ants seem ready to devote theirs deserve more than a scientific passing glance from us? Is that the stuff on which moral obligations are built?

And it does seem to me conceivable that some superbeing could, quite reasonably, take the same view of us. The human track record in matters cultural, intellectual, moral, and the rest, to which we attach so much value, could mean essentially nothing to the superbeing; and this not because of limitations in its intelligence and perception, but simply because by comparison with its accomplishments we just did not amount to anything. And if this were so, I do not see that we could mount a plausible argument to the beings in question for taking us under their wings (or whatever they had) and granting us rights and the rest of it.

In the absence of such an argument, there remains, of course, the option of philosophical name-calling. We can just insist that such beings are devoid of moral sense, Pure Practical Reason, or perhaps Humanity, among other things. In the latter case, at least, the tag would be literally justified, since the beings we are here contemplating would not literally be humans, whatever else they might be. But since we humans are not over-impressed at being accused of lacking Felinity, it is hard to see why our hypothetical ETs should be much impressed at being told that they lack Humanity. And as for lacking moral sense, they might well reply that they have perfectly good moral senses, and it is just that those senses do not tell them that they should be bothered about such insignificant creatures as humans. And it is unclear that Pure Practical Reason would require revision of that sense. Those who think otherwise, I submit, really do need to ask themselves how they stand *vis-à-vis* the ants.

One special category of cases to consider is that in which the super-beings, instead of simply having no use for us at all, quite definitely did have 'uses': wanted, say, to enslave people in various ways. Should we give in? Here we must bear in mind that in one sense, we would have no option: anything that the superbeings did not want us to do they could effectively prevent us from doing. But, as always with slavery, it does not follow that they could make us do everything, or even anything, they wanted. It would depend on what we were willing to put up with in the way of coercion: the prospect of death, for instance, or severe pain, or any number of other undesirable conditions could be threatened. But if we were willing to pay the price, we would not obey their commands. The situation might be comparable to that of a militarily weak country with a powerful neighbor equipped with nuclear weapons. Can the latter get the former to do anything it wants? Indeed not! The weaker group might even prefer general suicide to being subservient to the powerful country in question. The same might be true of us with respect to superbeings. So the question of whether we should give in is not an empty one, despite the overwhelming superiority of the beings we are considering. (I leave aside the possibility that the superbeings might re-wire us, thus turning 'us' into zombies or robots. They might do that, of course, but then we should no longer be able to face any questions about what to do, and thus would no longer be ourselves in any way relevant to moral questions.)

There is, then, a real question to be asked about this. But what is the answer? If another set of beings, invulnerable to our best efforts to destroy, damage, or injure them and overwhelmingly more powerful than us in every way, threatened us with various kinds of coercion which they were perfectly able to apply unless we did their bidding, should a moral human being do the bidding in question? I think it would be hard to give a general answer. It would surely depend on what they wanted us to do. It can hardly be true that no matter what it was, it would be beneath our dignity to do it. People subtly coerce their spouses and friends into doing things which both groups find quite pleasant in the event. And of course, sheer natural necessity makes us do many things which it can hardly be contrary to human dignity to do – eat, for example. *Prima facie*, it would seem that this is an area where we ought to let each person decide for himself where, if anywhere, he is going to draw the line; except, of course, in the case where the superbeings attempt to coerce us into treating each other immorally.

Subhumans

Alternatively, beings from elsewhere might prove to be markedly inferior to us in every capacity of interest. What then? The relevant analogy now would be to our duties towards the nonhuman animal world, the only differences being differences of species, or at any rate analogous to such differences. We would, in that case, have the usual scientific, collector's, and curiosity interests in those beings, and such interests may indeed create modest duties for us. But do they create duties to the beings themselves, as persons in their own right? Not obviously. The argument is simply the obverse of the foregoing, with ourselves in the superior rather than the inferior positions.

It is possible that the new creatures would be very cute, or otherwise attractive, quite apart from their inferiority in power. This would make a difference, just as it surely does in the case of dogs, cats, and human infants. It might make the difference that a general (but not universal) interest in the protection of those beings would arise. It also brings up a question about the notion of 'powers' that certainly requires some attention here. One thinks at first, of course, of such obvious things as ability to manipulate objects, to adjust means to ends, to employ abstractions, and so on. But 'cuteness,' after all, also denotes a sort of power – the power to amuse, even if not exercised with intent in the way that a professional comic does. And of course in that respect, and possibly some other ones peripheral to the notion of powers, the new beings might not be inferior to us. If this is accepted, then when the new beings had such characteristics we would not have the sort of case announced above, viz. of beings 'inferior to us in every capacity of interest.' How does this affect matters?

Typically, humans compare with each other as mixed cases. Few if any humans could claim to be 'markedly superior in every interesting respect' to any other human. And even if that were so, or nearly so, it would not follow at all that the one had no use for the other. Apart from the possibility of relationships of affection or friendship, it can, for example, pay *A* to hire *B* for a job that *A* could do better. *A* might have more profitable or interesting things to do with his time, while *B* might find the same activity pleasant or interesting as well as worth doing because of the amount *A* will pay him to do it. In economic terms, the situation will normally eventuate in incomes for both, *A* having the higher one. Whether extraterrestrials will figure in literally economic transactions will obviously depend very much on what specific properties the ETs have. But it is perfectly possible that even inferior ETs will have enough to offer to make it rational for us to carve out a moral niche for them, granting them rights of nonmolestation, and so on. Again, in short, it is not obvious that a fundamental revision of our moral ideas will necessarily ensue.

Very different nonhumans

We now come to what are potentially, I would think, the most interesting possibilities: creatures so different from us that it is difficult to compare their powers with ours – whose specific repertoire of capabilities overlaps ours very little, and in the regions of nonoverlap have nothing remotely similar to our own. Here if anywhere, one supposes, might lie the territory in which the seed of wholly new ethical concepts might germinate.

How might this go? Obviously a difficulty looms here: if the respects in which these beings differ from us were so different as to be inconceivable, then there is not much we could say. And in respects in which the differences are conceivable, a good many of them are surely likely to be, on the face of it at least, morally irrelevant. Now any difference one can think of could make a difference in the ways in which the being could be treated by us, and perhaps vice versa. Specifically, we can well imagine that a range of acts of ours might be harmful or beneficial to them while having the reverse effects on ourselves: water might be fatal to them, while at the same time bullets have no discernible effect. Or they might, despite having armor-plated hides, be just unbelievably thin-skinned, taking vast offence at remarks of ours that seemed utterly innocent. But the more fundamental principles of ethics, to the effect that one ought not gratuitously to harm others, including psychological 'harms' as coming under that rule's purview, would seem to provide the major premises for such cases (assuming we can make an argument for including the creatures sufficiently within the ambit of morality to make those premises applicable to them).

But suppose that the whole category of harms/benefits simply did not apply to the 'beings' with whom we are concerned. In that case, could moral notions be brought into play at all? Such might be the case with robots, for example. We take it that a robot has no feelings – a 'robot' with emotions would surely not be a robot any more. But if it has none, then what purchase could there be for a principle of respecting robots? It is, of course, possible to damage or destroy a robot, but we do not suppose that the robot itself has then been 'harmed' in any sense that is of direct moral interest. The true robot simply does not care whether it lives or dies. It does not even care whether it gives us the right answers: it simply gives us answers, but robots presumably have no preferences at all governing their behavior.

I am aware of such cases as those of Hal 9000 in the movie *2001*, or the two amiable androids in *Star Wars*, plainly manifesting preferences and emotions. But I am unconvinced. Either we think that those machines really were feeling those things, in which case we are inclined to say that they are, though not in physical respects, human; or, more likely, we suspect that they are not feeling those things at all but have been programmed to act as though they did. In so arguing, I am not insisting on any metaphysical impossibility of mechanical/electrical entities having feelings. I am, rather, insisting that those feelings matter for moral purposes, and that part of the current implication of the term 'robot' is that beings so-called lack genuine emotions and thoughts. Perhaps further developments will persuade us that this is a prejudice. They have not as yet persuaded me.

Some writers have recently professed concern for the rights of canyons, trees, and other insensient entities (Routley & Routley 1980). A theory of that kind could no doubt embrace robots and other beings devoid of the emotional or sensuous aspects of personhood – why not, after all? But such a maneuver, quite apart from its intrinsic implausibility, would obviously take us outside the realm of inter*personal* relations. We must then consider what sort of differences would be compatible with retaining the 'beings' with whom we are concerned in that realm – and yet make for significant departures from our basic moral ideas.

Let us review the classification produced on p. 250 above, with a view to considering which of these differences would matter, and how.

(1)   *Physical*
      One supposes that any number of physical differences would have substantial, even drastic impact on our usual affective responses to beings. And some physical differences would render ready communication with others exceedingly difficult, which in turn makes moral transactions problematic. But I see little prospect of sheer physical differences – independently of the other kinds of differences listed below,

which of course is not always possible – having much effect on our moral relations. Only if these differences would bring our beings within the purview of our previous discussions, by rendering them hopelessly inferior in powers or overwhelmingly superior, does it seem reasonable to suppose that they have any fundamental moral significance.

(2)   *Sensory*

Suppose that other beings are blind, but have fantastic radars; or are deaf, but are enormously sensitive to minute effects on ambient magnetic fields; or . . . . Again, some of these possibilities would affect the general repertoire of powers of the beings we are considering in such a way as to put them into the categories considered above. And clearly they will affect the particular rules of morality. It would be as wrong to inflict irreversible damage on the radars of our sightless but radar-equipped beings as it is to blind a fellow human, once the beings in question became otherwise eligible for moral status. Moreover, the sensory equipment of a being obviously affects our capability of carrying on effective communications with it. But it does so in a technologically, rather than a morally, relevant way. The being must be able to detect the signals we can send out, and to send out signals we can detect – if communication is to be possible. And, if it is not possible, it is difficult to see how moral relations can develop. Once it is possible, we would expect the range of communicational subject-matter between us to be greatly expanded. This will also undoubtedly alter the range of possible points of conflict, misunderstanding, or agreement and cooperation between ourselves and them as compared with what they were between humans and fellow humans. Still, these seem to be differences in content and not in form.

(3)   *Information-processing*

We are certainly prepared to envisage creatures whose calculational powers are enormously greater than ours, and probably also enormously different. (Could a creature be terrific at extracting cube roots, but scarcely able to add two-digit numbers? Or to recognize a good argument in *modus ponens*?) Obviously such capacities might be of great interest: imagine a Benthamite calculator, with a sensational ability to measure and integrate interpersonal cardinal utilities! But then, aside from the fact that such abilities well exceed what we would normally count as 'computational', it would again seem that the moral interest of the existence of such beings is essentially technical. Its remarkable capacities would qualify only as curiosities were we persuaded that utility is, after all, morally irrelevant.

A far more exotic supposition would involve our ETs in nonstandard logics and suchlike. For them, someone might suppose, $8 \times 2 = 23$, and 'P is not-P' expresses a truth. With such starting points, we may be sure, 'murder is right' will not be long in coming either. But we need not pause to consider such extravagances at any length. If we are to take unintelligible hypotheses seriously, why bother assessing any others for their plausibility?

It might be difficult, or at least extremely imprudent, to play poker with beings whose computational facility was far in excess of ours. But why should we think that anything very special in the way of moral notions should be appropriate in dealing with them? We do not currently think that mathematical geniuses have special moral status, for example. Nor do we think, for that matter, that computers have rights. It is difficult to see why we should bark up this particular tree in search of the morally special.

(4)   *Affective aspects*
Here we should perhaps make a distinction between two kinds of differences that might be comprehended under this heading. (*a*) The beings under consideration might display roughly the same general sorts of emotions that we do – love, envy, joy, anger, spite, and the rest – but they might display them in response to very different stimuli. Consider someone who sinks into the most profound depression upon seeing someone blink, though the death of a friend affects him not at all. It would be very difficult to deal with something like that, to be sure. (*b*) Alternatively, our creature might be subject to a *range* of emotions quite different from that of which we have any experience.

With the former kinds of beings, if the departures from the standard patterns were great, the problems posed in dealing with them would be great as well. With some emotions, at least, there is a question about the intelligibility of drastic deviation from the 'standard patterns:' is it possible to love someone for the very facts about him which make him a deadly enemy? Can anyone, however Martian, react with horror to an event which he regards as highly desirable in every possible respect? Nevertheless, there is certainly room for much variation, since even people we know vary a great deal in their emotional reactions. And again, this does not much affect the moral situation with respect to these varying persons. Content, yes, but evidently not form, is what seems to be affected here.

If the second possibility is realized to an appreciable degree – that is, not only do different situations generate recognizably similar emotions, but also their whole range of emotions is utterly different from ours – then it seems to me that very little can easily be said about how to relate to them. Among the principal means by which we understand other individuals is the sharing of emotions, or at least the appreciation, in the light of our own case, of how others must feel. But if we are wholly unable to feel what they feel, if their feelings bear no analogy, even, to ours, then it is hardly possible that sympathy or a sense of companionship could arise. And if those are impossible, then it would seem that the ambit of our moral relations to them must be correspondingly restricted.

Nevertheless, as with differences in values, as discussed next, one would think that some kind of accommodation of such emotions would be possible. We can learn to avoid those creatures when they are under the influence of these strange feelings. Perhaps more positive ways of coping with them could eventually be found. Otherwise, we just have to say that: whereof one can form no conception thereof one cannot speak.

(5)   *Values*

Here, certainly, we would expect to find divergences that matter. Now by 'values' here I do not mean, in the first instance, to restrict the discussion to moral values. Rather, I have in mind cultural values and what we might call 'life values:' what do these beings want out of life? At some point removed from what we normally understand of descriptions of such values, the beings will seem quite unintelligible to us. Picture some sort of Rube Goldberg device, one that apparently accomplishes nothing at all, but on the construction of which the creatures simply lavish care, attention, and ingenuity. Or imagine certain kinds of obscure movements which to us seem utterly meaningless but which to them have enormous significance. At some point, differences of this general sort could be radical enough to make communication with humans impossible.

But, within that very wide limit, do we not have a pretty good idea of how to deal with persons of differing values? So long as they are recognizably *persons*, anyway, one would think that treating them as such should be possible: we should be able to find areas in which they could pursue these deviant goals without running afoul of our very different pursuits. As with our fellow humans, we can assign rights, observe particular rules of etiquette and good form, and identify ways of helping or hindering their projects so that the moral virtues can be intelligibly displayed.

We can go farther than this, I think. For I would argue that the possession of values is a necessary condition for serious membership in the moral community. If a being had no values, no sense of what does and does not count in its life, then he would, I think, be unrecognizable as a person. Consider, for example, a being of fabulous cognitive capacity in some respects, and impressive strength or other physical abilities, but who quite evidently just does not care about anything (including, for instance, whether it lives or dies). It is difficult to conceive what such a being would be like unless it were wholly nonbiological in its physical constitution. And if it were, then are there not good examples of such entities right at hand in the form of digital computers? And does anyone seriously think that computers have rights, in and of themselves?

Conclusion

My argument in the foregoing will doubtless be found disappointing or deflating to those whose imaginations have roamed the limitless universe in search of the exotic; for my conclusion is that it is difficult to see how the possibility of nonterrestrial beings could very seriously affect the foundations of ethics, however drastically it might affect the details. Variation in the latter has not, I hope, been underplayed in this essay, and of course it must be agreed that we have no precise demarcation between what is 'detail' and what is 'fundamental.' But I have also contemplated the possibility of beings who, in one way or another, fail to qualify for moral consideration, either because they are far beyond us or because they are far beneath us. Some moralists would, I believe, want to contend that my account here is too niggardly, and to insist that moral duty reaches to every living thing. Here, I think, due consideration of the possibilities will encourage my less generous view of the range of duty. If a race of superbeings should move in, the survival of mankind on terms at all agreeable to us will, I fear, be a matter of sheer luck. It is when we contemplate the 'middle class' of extraterrestrials that we find, I think, interesting possibilities of variation in some of the details of ethics, while at the same time we remain convinced of the rightness of our basic moral principles.

Notes

1. I am freely paraphrasing here. The classic source for such principles is in Sidgwick (1907). But in fact, Sidgwick does not offer this principle. Instead he offers two others, which between them have the same effect. One is this: 'it cannot be right for *A* to treat *B* in a manner in which it would be wrong for *B* to treat *A*, merely on the ground that they are two different individuals, and without there being any difference between the natures or circumstances of the two which can be stated as a reasonable ground for difference of treatment' (*ibid.* p. 380). But this does not, as it stands, require anyone to *treat* other people equally; it merely requires that

either it is right for any two persons to treat people otherwise the same alike, or it is wrong for both of them – it cannot be right for one and not for the other to do whichever is in question. However, Sidgwick has another principle which he also thinks is self-evident: 'the good of any one individual is of no more importance, from the point of view (if I may say so) of the Universe, than the good of any other; unless, that is, there are special grounds for believing that more good is likely to be realized in the one case than in the other. And it is evident to me that as a rational being I am bound to aim at good generally . . .' (*ibid*. p. 382). He summarizes these, in effect, by suggesting that 'I ought not to prefer my own lesser good to the greater good of another' is 'self-evident' (*ibid*. p. 383). For much more on such matters, cf. Narveson (1984).

2. I owe the suggestion that ETs might display such nonindividualism to Edward Regis – as indeed I owe him thanks for the inspiration to compose this essay.

3. The classic discussion of this is found in Hume (1751), Section III ('Of Justice'), Part I: 'Let us suppose that nature has bestowed on the human race such profuse *abundance* of all *external* conveniences that, without any uncertainty in the event, without any care or industry on our part, every individual finds himself fully provided with whatever his most voracious appetites can want or luxurious imagination wish or desire . . . . It seems evident that in such a happy state . . . the cautious, jealous virtue of justice would never once have been dreamed of. For what purpose make a partition of goods where everyone has already more than enough? . . . Why call this object *mine* when, upon the seizing of it by another, I need but stretch out my hand to possess myself of what is equally valuable?' (*ibid*. p. 15.)

4. This has, indeed, already been argued, and by myself at that. In a portion of a lengthy critical notice of Nozick's *Anarchy, State and Utopia* (1974) deleted in the shorter published version (which is in *Dialogue*, 1977 (vol. XVI, No. 2)), I addressed myself to Nozick's defense of vegetarianism, pointed out that he appeared to think that individuals' rights are due to a property he refers to as 'M', which includes 'being a moral agent capable of guiding its behavior by moral principles and capable of engaging in mutual limitation of conduct; having a soul' (Nozick, 1974, p. 48). Nozick also conjured up the possibility of moral super-beings – e.g. on other planets – and asks whether we think it would be all right for ourselves to be sacrificed to serve the pleasure of such beings, implying that the answer being obviously in the negative, we ought not to suppose that the lower animals can be sacrificed to our pleasure either. But, I pointed out, the M-property neatly divides all creatures, with us on the winning side and animals, since they lack it, on the losing side. The argument in the present essay, however, abandons M-ness as a dividing line of that type. The view advanced here is that morality is a reasonable code of constraints on behavior for entities capable of adopting such codes to adopt in relation to whomsoever they may have to deal with. And it is not obviously reasonable for persons by comparison with whom we are ants – even though we are not actually ants – to constrain themselves in their dealings with us in ways that would only be appropriate if we were enormously more capable than we are.

References

Hume, D. (1751). *An Enquiry Concerning the Principles of Morals*. (Reprinted Indianapolis: Liberal Arts Press, Bobbs-Merrill, 1957.)

Kant, I. (1785). *Foundations of the Metaphysics of Morals*. (Transl. Lewis White Beck (1959). Indianapolis: Liberal Arts Press, Bobbs-Merrill.)

Narveson, J. (1984). The how and why of universalizability. In *Morality and Universality: Essays on Ethical Universalizability*, ed. N. Potter & M. Timmons. Dordrecht, Netherlands: Reidel.

Nozick, R. (1974). *Anarchy, State and Utopia*. New York: Basic Books.
Routley, R. & Routley, V. (1980). Human chauvinism and environmental ethics. In Mannison, McRobbie, and Routley, eds. *Environmental Philosophy*, Canberra: Australian National University, Department of Philosophy, Monograph Series No. 2, ed. D. S. Mannison, M. A. McRobbie and A. Routley, pp. 96–189.
Sidgwick, H. (1907). *The Methods of Ethics*, 7th edn. London: Macmillan. (Reprinted 1962.)

R. S. V. P. – A story

ROBERT NOZICK

The project began with high hopes, excitement even. Though people later came to think it just dumb, founded on a mistake so obvious that those who started it deserved its consequences, no one raised objections until well after the project was operating. True, everyone said it would be a long venture, probably not producing results for many generations. But at the beginning the newspapers carried frequent reports on its progress ('Nothing yet'). Practical jokers would call saying, 'Is this the Interstellar Communications Project? Well I'm a BEM you'd be interested in talking to,' or 'I have a collect call for the Interstellar Communications Projects from the constellation of Sagittarius. Will you accept the charges?' It was in the public eye, looked fondly upon.

Much thought had been given to deciding what listening devices to use and what sorts of signals to study intensively. What would be the most likely wavelengths for messages to come on? Would the messages be something like TV signals rather than consecutive prose? How would one tell that a signal was sent by intelligent beings rather than produced by some natural process? Investigating this last problem produced the Theological Project as a side-effect, for proponents of the argument from design, one traditional argument for the existence of God, had long wrestled with the same difficulties: couldn't any pattern, however intricate and wonderful, have been produced by some unknown mechanism? How could one be sure that an intelligence was behind it? Some foolproof test was needed, especially since, with a sufficiently complex manual of translation, any glop coming across could be decoded into an interesting message. Sending a return message and receiving a reply would take many years, perhaps generations, and it wouldn't do to have everyone on earth jumping for joy and holding their breath if they were just talking to the interstellar equivalent of the bed-post. The solution lay in abstract mathematical patterns, not realized (so far as anyone knew) in any actual causal mechanism and which (it was thought) couldn't be so realized. For example, there's no known causal process that generates the sequence of prime numbers in order; no process, that is, that wasn't expressly set up by an intelligent being for that purpose. There doesn't seem to be any *physical* significance to precisely that sequence, to a sequence which leaves out only the non-primes, and it's difficult to imagine some scientific law containing a variable ranging only over primes. Finding that a message began with groups of prime-numbered pulses, in order, would be a sure sign that an intelligence was its source. (Of course, something might be the product of an intelligent being even though it didn't exhibit such an abstract pattern. But a being wishing to be known to others would do well to include a pattern.) With alacrity, the theologians jumped on this idea, gaining their first National Science Foundation grant. Among themselves they called their project 'Hunting for God,' and the idea (about which other theologians had their reservations) was to look at the fundamental lineaments and structures of the universe, the clustering of galaxies, relationships among elementary particles, fundamental physical constants and their relationships, etc., searching for some abstract non-causal pattern. Were such a pattern found, one could conclude that a designing intelligence lay behind it. Of course, it had to be decided precisely which abstract patterns would count, and which features of the universe were fundamental enough. Discovering prime-numbered heaps of sand on some heretofore uninhabited island wouldn't do the trick, since one would expect to find something like that somewhere or other; what the significance would be of finding such patterns in cortical functioning or in the structure of DNA was a matter of dispute, with those viewing man as no more fundamental than the heaps of sand accusing their opponents of anthropocentrism. Theologians establishing the Reverend Thomas Bayes Society became expert in forming complex and intricate probability calculations and in debating delicate issues about assigning *a priori* probabilities. The results of the 'Hunting for God' project being well known, no more need be said here.

The initial excitement aroused by the Interstellar Communications Project was connected with a vague hope that other beings would enlighten people about the meaning and purpose of life, or with the hope that at least people would learn they weren't alone. (No one explained why the 'we' group wouldn't just expand, leaving people plus the others still quite alone.) After the project was set up, the best scientists went on to other more challenging tasks, leaving the rest to wait and listen. They listened and they examined and they computed and they waited. No qualifying abstract pattern was detected, nor was any message that looked intelligent even minus such a pattern. Since newsmen do not find a uniform diet of 'no progress' reinforcing, the project was reduced, in order to fill the auditorium for their third annual press conference, to inviting reporters from college newspapers, Sisterhood bulletins, and the like. Up gets this woman to ask why they should expect to hear anything; after all, they were only just listening and not doing any sending, why wouldn't everyone be doing the same?; maybe everyone else was just listening also and no one was sending any messages.

It is difficult to believe that the project had reached this point without anyone's having thought about why or whether extraterrestrial beings would want to try to make their presence known to others. Even though during the Congressional debate on the subject, in all the newspaper columns and editorials, no one once suggested setting up a transmitting station, no questioner asked whether other beings would do so. Little thought is required to realize that it would be dangerous simply to start sending out messages announcing one's existence. You don't know who or what is out there, who might come calling to enslave you, or eat you, or exhibit you, or experiment on you, or toy with you. Prudence dictates, at a minimum, listening in for a while to find out if other parties are safe and friendly, before making your presence known. Though if the other parties are at all clever, they would send reassuring messages whatever their intentions. They most certainly would not beam out TV signals showing themselves killing and eating various intelligently behaving foreigners. And if they're really clever, then (by hypothesis) they'll succeed in deceiving anyone not adhering to a policy of staying silent no matter what. Such considerations were neither explicitly formulated nor publicly expounded, but it must have been some feeling about the foolhardiness of broadcasting first (how else can one account for it?) that led to the notable but not-then-noted absence of proposals to establish broadcasting stations in addition to the listening posts.

Once again the project was a topic of conversation. 'Of course,' everyone said, 'it's ridiculous to expect anyone to broadcast: it's too dangerous. No interplanetary, interstellar, intergalactic civilization, however far advanced, will broadcast. For they don't know that an even more advanced and hostile civilization isn't lurking at the other end of their communications beam.' Interest in flying-saucer reports diminished considerably when the conclusion was drawn that the sending out of observation ships presents hazards similar to those of broadcasting messages, since the process of a ship's returning information to its source can be tracked. (Even if a ship were designed to give information to its makers by *not* transmitting any physical signal, or even returning to its base, there must be some contingencies under which it *would* do so, since nothing can be learned from a detection device that gives the same response no matter what it detects.) It was said that if its planning committee had included some psychologists or game-theorists or even kids from street gangs in addition to the scientists and engineers, the project never would have gotten started in the first place. The legislature wouldn't openly admit its blunder by ending the project completely. Instead they cut its funds. They did not authorize the broadcasting of messages. The members of the staff had various reasons for staying with the project ranging, one mordantly remarked, from masochism to catatonia. All in all, they found their jobs agreeable. Like night clerks in completely empty resort hotels, they read and thought and coped comfortably with the lack of outside stimulation. In that manner the project continued, serenely, for another eight years, with only a few comedians desperate for material giving it any mention at all; until the receipt of the first message.

Studious observation of reversals in public opinion and their accompanying commentaries has never been known to enhance anyone's respect for the public's intellectual integrity. (As for its intelligence, this would be a late date, indeed, to proclaim the news that the public adopts a view only after it is already known to be false or inadequate, or to note the general inability to distinguish between the first-person present tense of the verb 'to believe' and the verb 'to know.') People just refuse to admit that they have changed their minds, that they have made a mistake. So the very same people who said at first, 'How exciting, I wonder when the messages will begin arriving,' and who later said, 'How silly to listen for a message; it's too dangerous for anyone to broadcast,' now said, after the receipt of the first message, 'Of course a civilization *will* broadcast, even though it's dangerous, if it's even more dangerous for it not to broadcast.'

The first message picked up and decoded was a call for help. They were threatened by a coming supernova outburst of their star. No spaceships could escape the wide perimeter of destruction in time, and in any case they could not evacuate all of their population. Could anyone advise them about what to do, how to harness their star to prevent the outburst? Their astronomical observations had shown them that occasionally such outbursts didn't take place as predicted, and since they could discover no alternative explanation for this anomaly, they thought it possible that some civilizations had mastered a technique of inhibiting them. If no one told them how to do it, or came to their aid, they were doomed.

Over the next year and a half they beamed out their literature, their history, their accumulated wisdom, their jokes, their sage's sayings, their scientific theories, their hopes. Mankind was engulfed in this concentrated effulgence of a whole civilization, enthralled, purified, and ennobled. To many they became a model, an inspiration. Their products were treasured and they were loved. Did they view this outpouring as a gift to others, an inducement for others to help, a distillation for its own sake of the essence of themselves? No person knew or was prone to speculate as each, silently with them, awaited their tragedy. Never before had the whole of humanity been so greatly moved; never before had persons been so jointly elevated as in experiencing these beings.

At the end of a year and a half came a renewed call for aid; and in addition a call for some response, even from those lacking technical knowledge to help with the supernova. They wanted, they said, to know their messages had been received and understood, to know that what they held most important and dear would be preserved. They wanted to know, as they died, that others knew of them, that what they had done would continue, that it would not be as if they had never existed at all.

Only to the misanthropic can the ensuing debate have brought pleasure, the debate that raged among persons, and within some.

'It might be a trick, don't reply, it's too dangerous.'

'Beings capable of *that* civilization couldn't be up to trickery.'

'Perhaps they are quoting another civilization they've conquered, or an earlier phase of their own; Nazis could and would quote Goethe.'

'Even if they're not tricking us, perhaps some other aggressive civilization will overhear our message to them.'

'How can we let them perish without responding?'

'If we could help them escape their fate then certainly we should send a message telling them what to do, even though this would mean running serious risks. But we can't help them, and we shouldn't run risks merely in order to bid them a sentimental farewell.'

'We can save them from believing, as they die, that they are sinking into oblivion.'

'Why the irrational desire to leave a trace behind? What can that add to what they've already accomplished? If eventually the last living being in the universe dies, will that mean that the lives of all the rest have been meaningless? (Or is it vanishing without trace while others still remain on that is objectionable?)'

'How shall we face our children if we don't respond?'

'Will we have grandchildren to face if we do respond?'

No government sent a message. The United Nations issued a proclamation beginning with a lot of 'whereas's' but containing near the close a gathering of 'inasmuch's' so it didn't proclaim its proclamation of regret very loudly. But it did issue an order, in its stated role as guardian of the interests of the earth as a whole, that no one endanger the others by replying. Some disobeyed, using makeshift transmitters, but these were seized quickly, and their signals were too weak to reach their destination intact through the interstellar noise.

Thus began the grim watch and countdown. Watching for their rescue, listening for some word to them from elsewhere. Waiting for their doom. The time, for which their astronomers and earth's also had predicted the supernova outburst, arrived. Some persons paused, some prayed, some wept. All waited, still.

The existence of a finite limit to the velocity of causal signals had been of some interest to physicists. Epistemologists had worried their little heads over the question of whether what is seen must be simultaneous with the seeing of it, or whether people can see far into the past. Now came the turn of the rest. The fate of that distant planet was already settled, one way or the other, but knowledge of it was not. So the wait continued.

For another year and a quarter, remembering their debates, mulling over their actions and inactions, contemplating the universe, and themselves, and the others, people waited. Poetically just things could have happened. A message could have arrived saying it was humanity that had been tested, it was the sun that was due to outburst, and since the earth's people hadn't ventured to render others aid or comfort, others would not help them. Or, they could have been rescued. (How greatly relieved people then would have felt about themselves. Yet why should someone else's later acts so alter one's feelings about one's own?) But the universe, it would appear, is not a poem. No messages to them were detected. Light from the outbursting of their star reached earth as their broadcasts (should they have terminated them a year and a quarter before the end?), as their broadcasts and their plays and their science and their philosphy, their hopes and their fears and their courage and their living glow ended.

Some people used to think it would be terrible to discover that human beings were the only intelligent beings in the universe, because this would lead to feelings of loneliness on a cosmic scale. Others used to think that discovering intelligent beings elsewhere would remove their own last trace of uniqueness and make them feel insignificant. No one, it seems, had ever speculated on how it would feel to allow another civilization to vanish feeling lonely, insignificant, abandoned. No one had described the horrendousness of realizing that the surrounding civilizations are like one's own; of realizing that each neighbor remaining in the universe, each of the only other ones there are, is a mute cold wall. Limitless emitlessness. Lacking even the comfort of deserving better, facing an inhabited void.

INDEX

Albertus Magnus, 5
Alpha Centaurians, 232, 249
anticryptology, 202
Apollo program (NASA), 139
Aquinas, St Thomas, 5
Arecibo interstellar message, 80
Arecibo, planetary radar, 186
Aristotle, 3, 4, 43
arithmetic: and communication, 122; and
    efficiency, 121
artificial intelligence, 19, 81; see also
    Minsky, Marvin
Astrosearch, 181
Augustine, St., 5
Australopithecus africanus, 51
Ayala, F. J., 31, 133

Ball, J. A., 98
Barnard's star, 8
Bayes, Thomas, 268
Beck, Lewis White, 1, 44
Black, David, 176
Bobrow, Daniel, 120
Bowyer, Stuart, 184
Bracewell, R. N., 143
Bradbury, Ray, 107
Brewster, Sir David, 45
British Interplanetary Society, 138
Brown, S., 79
Bruno, Giordano, 3, 5

C3P0, 53
Callimahos, Lambros, 204
Cambrian era, 26, 33, 49
Campanella, 6
Canadian National Museum, 36
categorical imperative, 64–5, 248–9
causes and goals, 125–7
Chalmers, Bishop Thomas, 6
Chardin, Teilhard de, 98

Chewbacca, 53
civilization: and communication, 29;
    lifetime of technical, 158–9; and
    technology, 28
Clark, Tom, 176, 177
Cocconi, G., 173, 231
Cohen, Chip, 183
colonization: galactic, 153–6; of space,
    135–40
communication: and arithmetic, 122; and
    cryptology, 201–14, see also cryptology;
    encoding, 165; human, 127; motivation
    for interstellar, 142; pheromones, 51
computers: and space exploration, 135
Connes, P. 170
Copernican revolution, 5, 15, 236
Copernicus, 5
cosmic haystack, 185–6
Cosmic Search, 180
Cosmos (Sagan), 131
Crick, F. C., 130
cryptology: decipherment, 202; message
    recognition, 202; nature of signal, 203;
    randomness, 210; signal analysis, 205–10
Cuzzi, Jeff, 176
Cyclops study, 134, 179

Darwin, Charles, 7, 61; Descent of Man,
    61, 171
Darwinian revolution, 5, 15, 21
Darwinism Defended (Ruse), 21
Deavours, Cipher A., 164, 236
Deep Space Network (NASA), 184
Descartes, R., 11, 59
Descent of Man (Darwin), 61, 171
design, argument from, 4, 13, 267
determinism: and science, 25
Dilthey, William, 91
dinosaurs, extinction of, 34
Dixon, Robert, 180

DNA, 130; recombinant, 145, 153
Dobzhansky, T., 31, 133
Dole, S. H., 96, 98
Donne, John, 111
Drake equation, 140–2
Drake, Frank, 129, 133, 174–5, 231, 234;
   Project Ozma, 164
Dyson, Freeman, 130, 133, 146, 187, 233

economics: principle of, 117
Eddington, Sir Arthur, 98
Efflesburg telescope, 183
Eigen, Manfred, 20, 24, 47
Eisley, Loren, 113, 136
Epsilon Eridani, 174
eschatology, 13
Esperanto, 222
E.T., 28, 43, 44, 49, 51, 52
Euclid, 224
evolution: of alien intelligence, 50, 133;
   and chance, 27; of complex life on
   Earth, 32; convergent 20–1, 35–8; and
   determinism, 34, 38; extraterrestrial, 10,
   46–8; eukaryotes, 26; of the eye, 28; and
   grammar, 123–4; of *Homo sapiens*, 31,
   34; of independent humanoid, 36; of
   intelligence, 28; of intelligence in
   extraterrestrials, 50, 133; kin selection
   61–2; likelihood of intelligence, 99;
   mental 118; natural selection, 47, 118;
   physical and mental, 52; prokaryotes,
   26; rate of, 33; science as a result of
   100–2; value of intelligence and, 99; *see
   also* extraterrestrials, evolution of
evolutionary convergence, 20–1, 35–8
exobiology, 7, 8, 14
exo-sciences, 14
exo-sociology, 14
extraterrestrial intelligence:
   anthropomorphic nature of 10;
   anthropomorphism, dangers of, 12; and
   arithmetic, 122; belief in, 3, 147; belief
   in, ancients', 4; civilization of, 14;
   cognitive anthropomorphism, 113;
   consequences of contact with, 267–73;
   communication from, 10, 11, 12, 20, 23,
   81, 163; ethics of, 240; evolution of, 50,
   133, 140; existence of 24–5; likelihood of
   communication with, 134; like *Homo
   sapiens*, 35; mathematics of, 81; moral
   relations with, 248–63; nature of, 21;
   non-existence of, 133–50; science of, 80,
   83–114; space exploration of, 135;
   speculation about, 13; as subhumans,
   257–8; superhumans, 252–7; synchrony,
   10; and technology, 28–9, 38, 100; as
   very different non-humans, 258–63;
   visiting earth, 9

extraterrestrials, consciousness and, 52–4;
   communication with, 55; human moral
   obligations to, 68–71; morality of, 60–5,
   68; sexuality of, 49

Face that is in the Orb of the Moon
   (Plutarch), 4
Fermi, Enrico, 129, 133
Fontenelle, Bernard de, 6
fossil record, 32, 109
Francois, J., 133
Frautschi, S., 158
Freitas, Robert, 179
Freudenthal, Hans, 127, 165, 203
Friedmann, William F., 201
*From the Closed World to the Infinite
   Universe* (Koyré), 111

Gassendi, 3
Gauss, Karl Friedrich, 121, 170
General Problem Solver (GPS), 127
Gold, Thomas, 177, 183
Gould, J. L., 40
Great Moon Hoax, The (Locke), 15
greatest happiness principle, 64–5, 69, 72
Growth of Biological Thought (Mayr), 20
Gulkis, Sam, 184

Hal 9000, 259
Hart, Michael, 129, 133
Homo habilis, 51
Homo sapiens, 27
Horowitz, Paul, 181
Hubble, Edwin, 171
human rights, 235
Hume, David, 60
Hunter, M. W. III, 137
Huxley, Thomas Henry, 7
Huygens, Christiaan, 6, 92, 112–13

Immense Journey, The (Eisley), 113
Infrared Astronomical Satellite (IRAS),
   187–8
intelligence: definition of, 118; and
   problem solving, 118–19; non-human, 31
intelligent behaviour, 39

James, William, 93, 95
Jansky, Karl, 164
Johnson, Samuel, 44
Jones, Eric M., 154

Kant, Immanuel, 3, 43, 62, 64, 85, 113,
   248–9
Kant–Laplace hypothesis, 7, 8
Kappa test, 206–7, 211–12
Kober, Alice, 204
Koyré, Alexandre, 111

Krauss, John, 180
Kuhn, Thomas, 94
Kuiper, T. B. H., 133, 146

Lambert, 3
Leia, Princess, 53
life: advanced, 9; carbon-based, 19, 113; combinatorial problem, 20; conditions for, 8, 9, 32; definition of, 19; extraterrestrial, 19, 21–4, 44–6, *see also* extraterrestrial intelligence; in extreme environments, 33; multiple origins of, 32; origin of, 23, 47; planets capable of supporting, 96–9; primitive, 9; probability of extraterrestrial, 33; silicon-based, 19; spontaneous generation of, 8
LINCOS (Lingua Cosmica), 127, 165, 203, 215–28
Linear B, 204
Locke, John, 3
Locke, Richard Adams, 15
Lowell, Percival, 144
Lucian, 6
Lucretius, 4–8

MacArthur–Wilson theory, 140
MacGowan, R. A., 98
Marconi, Guglielmo, 171
Mars, 'canals' of, 7; intelligence on, 144; missions, 23; life on, 233–4
Mayr, Ernst, 20, 31, 133
memory, holographic, 125
Menzel, Donald, 23, 25
Microwave Observing Program (NASA), 185
migration, interstellar, 129, 131
Miller, Stanley, 47, 121
Milton, John, 6
Minsky, Marvin, 81, 165
monism, 93, 112
Montaigne, 6
morality, subjective elements in, 63
moral relevance, 246–52
More Worlds than One (Brewster), 45
Morris, M., 133, 146
Morrison, Philip, 37, 133, 143, 173, 231

Narveson, Jan, 229–30
NASA, 37
National Academy of Sciences, 3, 171
National Radioastronomy Observatory (NRAO), 174, 176
natural selection, 39
Neanderthal Man, 28
Newell, A., 127
Newman, William I., 131
Nicholas of Cusa, 3, 5

Nozick, Robert, 229–30

Ohio State–Wesleyan University Radio Observatory (OSURO), 180, 181
Oliver, Bernard, 212
O'Neill colony, 136, 146
Ordway, F. I., 98

Paine, Thomas, 6
Painvain, Georges, 209
Paleozoic, 26
Palmer, Pat, 177
Pascal, Blaise, 13, 111, 112
Pauli, Wolfgang, 121
Pearman, J. P. T., 98, 99
Pickering, W. H., 144
Pioneer plaques, 80
Planetarians, 92
Planetary Society, 181–2, 185
planets, life-bearing, 19
pluralism, 93, 112
plurality of worlds, 5, 6
Plutarch, 4, 5, 7
Pope, Alexander, 6
Precambrian, 33
prime numbers, 203, 268
Principles of Paleontology (Raup and Stanley), 21
Project Daedalus, 138, 139
Project Ozma, 164, 174–5
Putnam, Hilary, 57

Quine, W., 12

R2D2, 53
Raup, David M., 20, 79, 238
Regis, Edward Jr, 229–30
Rerum Natura (Lucretius), 4
Rescher, Nicholas, 80, 242, 250
Return of the Jedi, 71
rights, 255, 262
RNA, 47, 130
Rood, R., 239
Rosetta Stone, 204
Ruse, Michael, 20
Russell, Dale, 36

Sagan, Carl, 96, 102, 131, 133, 142, 144, 229–30, 233–4, 236–7, 239–40
Salvini-Plawen, L., 28
science, cognition and, 94–5; comparability of ETI's and ours, 103–5; conceptualization of, 87; conditions for development of, 107; diversity of, 84–9; extraterrestrial, 83–114; ideational innovation in, 87; machinery of formulation, 85; orientation of, 85; significance of, 90; topics of, 86

Sentinel, 181–2
Serendip, 184
SETI (Search for Extraterrestrial
  Intelligence): AMSETI (Amateur
  SETI), 181; argument for, 161;
  consequences of contact, 15, 231, 236,
  267–73; consequences of non-contact,
  233; Cyclops, 79; decipherability of
  transmissions, 9, 164, 201–14; dedicated
  searches, 179–83; detectability without
  intelligence, 41; and evolutionary
  convergence, 36; false alarms, 214;
  financing, 243; forged messages, 213–14;
  magic frequency, 182; NASA proposal,
  134; parasitic searches, 183–5;
  philosophical issues and, 1; program,
  36–7; programs 1959–84, 191–9; project,
  23; significance of contact, 229–30;
  strategy, 31, 42, 167–89; Suitcase SETI,
  182; waste of money, 29; what to search
  for, 170–3; worth doing, 14
Shapley–Curtis debate, 171
Shklvoskii, I. S., 102
Simmel, Georg, 94
Simon, H. A., 127
Simpson, G. G., 20, 29, 31, 35, 109, 133,
  242
Singer, Peter, 69
Skywalker, Luke, 53, 70–1
space travel, 9, 135–40
sparseness principle, 117, 119–23, 127
spontaneous generation, 7
Stanley, S. M., 21
Star Wars, 53, 70, 245, 259
Sullivan, W. T., 79
Superman, 44
Swift, Jonathan, 6
Systematics and the Origin of Species
  (Mayr), 20

2001, 259
Tarter, Jill C., 129, 164

Tau Ceti, 174
Tertiary, 27
Tesla, Nikola, 170
Tipler, Frank, 130; solipsistic world view
  of, 152–3
Trefil, J., 239
Turing, Alan, 120
Turing machines, 120

ultimate reality, extraterrestrials and,
  56–60
universal knowledge, 54–56
universe, as finite, 111
universe, intelligence in, 147
Urey, Harold, 121

Vader, Darth, 70–1
Valdes, Frank, 179
Ventris, Michael, 204
Viking Mars mission, 233–4
Voltaire, 6
von Littrow, Joseph, 170
von Neumann, John, 130
von Neumann machines, 130, 131, 152–3,
  238; as interstellar probes, 135–47
Voyager Interstellar Record, 80
Voynich manuscript, 203, 236

Wallace, Alfred Russel, 7
Watson, James D., 130
Wells, Orson, 15
Wetherill, C., 79
Wheeler, J. A., 147
Whewell, William, 3, 43
Wilkins, Thomas, 6
Wilson, E. O., 51
Winch, Peter, 14

Yoda, 44, 49, 53

Zuckerman, Benjamin, 177